Introduction to chemical ecology

Introduction to chemical ecology

Michel Barbier
Institut de Chimie des Substances Naturelles, CNRS
Gif-sur-Yvette, France

Translated by M. Ferenczi

Longman
London and New York

Longman Group Limited London

*Associated companies, branches and representatives
throughout the world*

*Published in the United States of America
by Longman Inc., New York*

© Masson, Editeur, Paris, 1976
English translation © Longman Group Ltd. 1979

First published in French under the title Introduction à l'Écologie Chimique.
This English translation first published in 1979.

British Library Cataloguing in Publication Data
Barbier, Michel
  Introduction to chemical ecology.
  1. Ecology   2. Biological chemistry
  3. Environmental chemistry
  I. Title
  574.5   QH541   78-41033

  ISBN 0-582-44378-4

Printed in Great Britain by
Richard Clay (The Chaucer Press) Ltd, Bungay, Suffolk

Contents

Foreword

Michel Barbier, the author of this book, is a chemist, an organic chemist – master of all the experimental methods of modern organic chemistry which he puts to the service of descriptive biochemistry. By choice, he studies the molecules which are specific to invertebrates, at the Institut de Chimie des Substances Naturelles of the CNRS at Gif-sur-Yvette, France.

Does this work qualify him to produce a book of this kind on *Ecology*, even within the bounds implied by the title *Chemical Ecology*? I believe it does, and this belief can be justified through arguments taken from the history of science. I have in mind particularly two of the most famous chemists, who played a predominant role in what was to be known later as ecology. I will not burden Michel Barbier by a comparison with Lavoisier and Pasteur; it is already quite a compliment to say that he is worthy of being placed in the wake of these great men.

We would be far from grasping the importance of the work of either of these men of genius if we were to believe that the possession of the facts on which their work was built rested on chance alone. Neither the work of Lavoisier nor that of Pasteur owes anything to such circumstances. It was built up by slow persevering effort always faithful to itself. This effort helps to reinforce through experimentation the ideas which strengthen the inductions brought about by reflection. Both these scientists remained constantly aware of a very general idea, perhaps one of the most general ideas: the idea of the unity of science, according to which the natural laws are universally applicable.

The concept of determinism had not evolved in their day; this happened later under the powerful influence of the thought of Claude Bernard. It already existed, however, in a vague and rudimentary form and it is indeed this elementary concept of determinism, which was rejected by most biologists and all doctors when they espoused the idea of the spontaneity of living matter, the reactions of which could not be predicted, in principle, and not because of a lack of understanding which might have been considered to be temporary.

Even in the eyes of every schoolchild, Lavoisier is thought of as the founder of modern chemistry. The more learned cannot deny that he was the genuine ancestor of biochemists and physiologists. A short while before being guillotined, he produced a very short text of an extraordinary density concerning the 'movement of matter on the surface of the globe', which I believe can be considered as the origin of ecology. Even Pasteur himself did not disdain to stress its relevance. In spite of this first class recommendation, this text still remains almost forgotten. Here are the main sections which are well worth reading over and over again by anyone concerned with what was known, at the beginning of the century, as *Natural Philosophy*:

Plants extract from the ambient air, water and generally from the mineral kingdom the substances which they require for their organization.

Animals either feed on plants or on other animals which live off plants in such a way that the substances they are made of always are, in the final analysis, either extracted from air or from the mineral kingdom.

In the end, fermentation, putrefaction and combustion always give back, either to the atmospheric air or to the mineral kingdom, the principles which plants or animals had borrowed for a time. By which means does nature achieve this prodigious circulation between the three kingdoms?

This kind of scientific will bequeathed by Lavoisier which was remarkably ahead of its time had only just been rediscovered by Jean-Baptiste Dumas when Pasteur used it as his theme in April 1862 for his progress report to the Minister of Public Education about the state of scientific knowledge in his own field. In those days, it was customary to assess in this way the state of scientific progress and Pasteur chose this method to disclose the 'role of fermentation in nature' and showed that his research work fitted into the mainstream of the 'Lavoisier' tradition.

The words of Lavoisier, said Pasteur, *present, with admirable clarity, the three facets of the fundamental question of the perpetuity of life at the surface of the earth. The first two form the object of modern physiological research. The third one, namely, to use Lavoisier's words, 'the incessant return of the principles borrowed by plants and animals to atmospheric air and to the mineral kingdom', has hardly ever been touched upon.*

I am close to being embarrassed to have to admit that this is precisely the field which I pretend to have taken up. At least, I concentrated my efforts overy many years to bring some light to it by direct experimentation and I am happy to have obtained some results. First of all, I would like to emphasize to a certain extent the relevance of this work.

It is widely known that substances extracted from plants will ferment when left to themselves and will slowly disappear when in contact with air. It is also known that animal carcasses putrefy and that soon enough only their skeleton remains. These destructions of dead organic matter are one of the requirements for the perpetuity of life. If the remains of dead plants and animals were not to be destroyed, the globe's surface would be thronged with organic matter and life would become impossible because, carrying further the idea of Lavoisier I borrowed earlier, the cycle of transformations could not be completed. In other words, if in a living being the internal movements which regulate the laws of life were to stop, the death process would only be starting. For it to come to its end, it is necessary that the organic matter of the dead body, whether animal or plant, is returned to the simplicity of mineral combinations.

It is necessary for the fibrin of our bones, urea of urine, the ligneous parts of plants, the sugars of their fruits, the starch of their seeds, etc. to be slowly reduced to the state of water, ammonia and carbonic acid so that the elementary principles of these complex organic substances can be again taken up by plants, rebuilt into more complex structures and used again as food by new beings, similar to those which were responsible for their birth, and so on in saecula saeculorum.

How are all these transformations achieved? This is a problem which can be subdivided into a multitude of others, all very interesting and full of prospects, and to which I am searching for a solution. I have already contributed six years of the most intense work to this study and it seems to me that I can say with confidence that my first results allow me to perceive already the more general law governing this type of phenomenon. I am brought to the conclusion that the destruction of organic matter is essentially due to the multiplication of microscopic, organized beings which are in possession of special abilities enabling them to break down complex organic substances by way of slow combustion processes and oxygen fixation. These properties render these agents the most active factors contributing to this necessary return to the atmosphere of everything that was living, which I have mentioned above.

I demonstrated that the atmosphere in which we live ceaselessly carries the germs for these microscopic beings which are always ready to multiply on dead matter, in order to play the destructive role which is the correlative of their own life. And if the organic laws which govern the changes in the tissues and liquids of animal bodies did not prevent the propagation of these micro-organisms, at least under the normal conditions of life and health, we would at every instant be liable to their invasion. But as soon as the breath of life expires, there is not a single part of the organism, whether animal or vegetable, which is safe from being used as food. In short, after death, life reappears under a different form and with new properties. The germs of these microscopic organisms which are to be found everywhere start their evolution and under their influence, organic matter is sometimes transformed into gases by fermentation processes and sometimes atmospheric oxygen is absorbed in very large quantities, and little by little achieves its complete combustion.

If we can imagine it at all it is mind-boggling to consider the gigantic mass of sugar compounds which are accumulated by nature every year in all plant life at the surface of the earth. It is absolutely necessary for these millions of tonnes of sugar to be destroyed and returned to the atmosphere. Even if man who utilizes some of this sugar in his food intake did not burn up any through respiration, the complete combustion of these immeasurable quantities of sugar would nevertheless still exist, because, and I would like to emphasize this point, it is written in the laws of the perpetuity of life on earth that every single part of a plant or animal must be destroyed and must be transformed into volatile, gaseous or mineral substances.

What are the mechanisms by which nature is able to destroy these unfathomable quantities of sugar compounds which every year are synthesized by the plant kingdom? As soon as even a tiny droplet of a sweet juice is left to itself, air brings to it the germ of a small mycodermic plant which will propagate with astonishing ease and, with a direct correlation to the life and multiplication of this organism, the sugar will be transformed into alcohol and carbonic acid. This small plant is one of the many organized sugar ferments. We can see that in the first stage of these phenomena, sugars are, partly, beginning to be returned to the atmosphere since one of the principles resulting from its decomposition is carbonic acid gas. However, alcohol still remains to be destroyed. I have recently established with complete certainty that alcohol is destroyed when under the influence of a microscopic plant, albeit different from the plant mentioned above. Its germ is also carried by air and in the new alcoholic liquid, this plant possesses the very remarkable ability to use up atmospheric oxygen and to combine it with alcohol to transform the latter into acetic acid. Then, if this microscopic plant is given the opportunity to carry on with its work, oxidation processes will in time be extended to acetic acid itself which will be totally transformed into carbonic acid and water, which are the end products of the total destruction of sugars and the realization of their complete return to air and to the atmosphere.

Let Michel Barbier, let the readers of this book forget these long quotations: I am but an old man for whom nothing remains of the eagerness which he had while young to push himself forward, and who utilizes whatever influence he might still carry to divulge the work of these giants who were our predecessors.

These are the reasons why I wanted to give to my young friend Michel Barbier the patronage of these great minds. The summary of the first important discoveries in the fields of biochemistry and microbiology of Pasteur, which are placed by their author in the perspective of the questions formulated by Lavoisier some 70 years earlier, is it not one of the most beautiful examples of the sort of calls which, throughout the centuries, men of genius send to each other? I will take this opportunity to repeat the sublime verse at the end of the poem 'Les Phares' by Baudelaire:

Car c'est vraiment, Seigneur, le meilleur témoignage
Que nous puissions donner de notre dignité
Que cet ardent sanglot qui roule d'âge en âge
Et vient mourir aux bords de votre éternité!

Ernest Kahane

Preface

All things are born out of struggle and necessity.

(Heraclitus)

Chemical ecology is the science of the chemical relationships between living organisms, or between the living and the mineral world. These interactions are characteristics of life itself. Consequently they represent an infinite and highly complex field of investigation. For many species, the real problem is not merely to find a means of improving life, but rather to find ways of surviving at all. The critical aspect of life therefore becomes the struggle to adapt, or attempts at adapting to, the surrounding environment. Active molecules can be the agents for the transmission of categorical information which must be translated into its original form by specialized receptors such as in the case of pheromones. In other instances (defence substances) these molecules act directly. Furthermore, these compounds can also correspond to specific biological requirements (sterols). The progressive degradation of the biosphere and the threat which man represents to his environment and consequently to his future drew attention to the problem of the need to protect and understand. This change in attitude led to motivation, including the search for a better understanding of nature. Chemical ecology is one of the keys which helps us attain such knowledge.

This work is presented in the form of an introduction, with no other aim than to describe examples of the main types of interactions. It was first written as a course of lectures which were given in Paris and later in Rio de Janeiro, and the text was then further developed and improved. The bibliography is limited to the principal publications with an emphasis on books and reviews. A spectrum of examples has been chosen which ranges from phytotoxins, mycotoxins, antibiotics to defence chemicals, pheromones, etc. The book ends with an example of pollution: the intoxication of the oceanic environment.

In short, the book is directed towards all those who would find useful a synthesis of the elementary concepts of the subject illustrated by the more striking examples, in addition to a series of references which provide the reader with the key to more technical publications.

Chapter 1

The origins of, and introduction to, the concept of chemical interaction in nature

Chemical ecology

It is only now, at the end of the twentieth century, that man has come to realize that any attempt at predicting his future must be preceded by a modification of his preconceived ideas about nature. In showing the identity of water with life Cousteau explained, in 1970, the oneness of pollutions: there is only one pollution, and it is synonymous with death. It is the problems raised by pollution which have forced man to adopt a common language, namely that of the absolute necessity to protect the biosphere. We have, therefore, to reconsider nature: we cannot keep on ignoring our environment.

Behind the magic of words loaded with emotional meaning and behind the fear of irreversible losses, the sad reality of destroyed equilibria is apparent. Natural equilibria are ceaselessly destroyed but constantly re-established and are parts of dynamic systems, of natural cycles. The factor that distinguishes man's effect on nature, is the irreversible aspect of most of his destruction. Yesterday's naturalist was merely curious; today he must also protect. The observer has been able in the end to predict the evolution of ecosystems. Entomologists and botanists have had to change the dimensions of their world and to become ecologists.

Since the time of the Renaissance, the whole of nature has become the object of physical science; after the necessary work of classification, natural science has aimed at establishing rational principles. However, the concept of nature continued to oscillate between vitalism and rationalism. The forefathers of ecology were two chemists, Lavoisier and Pasteur and, since their work, ecology and chemistry have become irrevocably linked. Lavoisier had a clear idea of the role of the circulation of the elements in the living world, as can be seen in a note published in the *Mémoires de l'Académie des Sciences* in 1792: 'Fermentation, putrefaction and combustion continuously give back to the air of the atmosphere and to the mineral kingdom the principles which have been borrowed by plants and animals. By which means does nature achieve this marvellous circulation between the three kingdoms? . . . since combustion and putrefaction are the means used by nature to give back to the mineral kingdom the materials which had been extracted to form plants and animals, vegetation and animalization must be the inverse phenomena of combustion and putrefaction.'

This text by Lavoisier (Kahane, 1974) is remarkable not only because it establishes the fundamentals of ecology, but also because it heralds Pasteur.

More recently, in a discussion prompted by remarks of Claude Bernard on the difficulty, indeed the futility, of trying to define life, Oparin strongly argued in favour of a common link between all living beings: the interaction between organisms and their environment: 'This specificity indicates by a red thread all the line of life.' Recent developments in the understanding of chemical interactions between living organisms bring a fundamental contribution to rational understanding. Furthermore, they lie at the starting point of any action towards environment protection.

Life appeared on earth approximately three thousand million years ago. The timespan in which we live contains the outcome of the ruthless necessity for the adaptation of populations to the changing conditions of their environment. The occurrence of mutants created modifications from which fantasy was excluded inasmuch as it was no part of well-adapted entities. The resulting whole is amazingly complex. Evolution is complication. This complexity, however, gives the impression that a certain amount of order is laboriously and progressively being introduced amidst total chaos. This is merely a subjective observation and one must avoid finalistic arguments which such an observation can lead to. 'Given time, everything possible would happen' (Heroditus). 'Given time, only what is likely would occur' (Kahane, 1973). Given time, only what is necessary survives. The application of the requirements of living matter to the results of probabilities has led to the progressive adaptation which we refer to as evolution.

Living organisms of both kingdoms interact with the immediate environment by means of chemical molecules; animals affect animals or plants, plants affect plants or animals. In addition, the effect of the mineral kingdom on both plants and animals should be considered. The study of these interactions and the study of the chemicals responsible for these interactions is the domain of chemical ecology (Sondheimer & Simeone, 1970).

In the light of recent developments, man's understanding of nature was continuously being modified: new concepts and new words were coined. Chemical ecology appears to be concerned with the need to reconsider the understanding of nature. We know that, by definition, ecology is the science concerned with relationships between the organism and its environment. A long-standing classification scheme distinguished between autecology, or ecology in the narrow sense of the word, and demecology or the dynamics of populations, and also synecology which concerns the interactions between organisms themselves. Since synecology only deals with interspecific relations, chemical ecology seems to be more closely related to the word biocoenotic, which refers to inter- or intraspecific relationships.

The rapid improvement in man's knowledge of this field is due to the development of techniques, particularly since 1950. Indeed, because most interactions involve a large number of very small amounts of substances, only modern analytical techniques have made these studies possible. Nuclear magnetic resonance spectrometry linked to Fourier transform analysis, mass spectrometry and X-ray crystallography, enable one to study chemical structures using less than a milligram of material. The basic concepts of chemical ecology have been adequately stated by Florkin in 1966, who at the same time introduced a vocabulary and a system of definitions that are most interesting and which we will consider in greater detail later. In Florkin's chapter concerning the biochemical continuum, the main ideas and the web of present day chemical ecology are encountered: 'It clearly appears that in the network of the biochemical continuum, a flow of specific molecules or of macromolecules which carry a certain quantity of information is taking place.'

Chemical interactions are not restricted to the transmission of coded messages. At every level in the evolutionary scale, chemical weapons are being used, both for offensive and defensive purposes. Very early on man had to learn about the existence of venomous animals and poisonous plants. The numbers of contemporary deaths caused by the bites of snakes, scorpions and spiders or by the consumption of *Amanita phalloides* are still far too high.

The complexity of relationships between organisms led authors to develop a language and classifications. The first consideration is the inter- or intraspecific character of the relationship followed by the resulting effects: whether beneficial or detrimental to the species responsible for it (positive or negative evolutionary effect). Table 1 summarizes the classification suggested in 1971 by Whittaker and Feeny. Among interspecific interactions (allelochemical interactions, Whittaker 1970 a and b) allomones are found (Brown *et al.,* 1970) which result in an advantage for the species producing them (repellents, defence chemicals, venoms, attractants, etc.) and kairomones which are advantageous to the organisms receiving them (growth factors, warning signals, chemicals enabling host–plant relationships, etc.). Among the various types of intraspecific interactions, are found the various types of pheromones as well as growth inhibitors and autotoxins.

Table 1 also indicates the importance of pollution phenomena which are found mainly under two headings. Depressors are thought of as waste products which poison the receiving organism without being useful to the producing organism. These chemicals are part of the interspecific effects. Autotoxins which are classified among intraspecific effects are toxic for the organism which produces them without conferring any evolutionary advantage to the organisms

Table 1 Classification of the various types of interactions according to Whittaker and Feeny (1971)

I. *Interspecific chemical effects (or allelochemical effects)*
A. *Allomones:* adaptive advantage for the producing organism.
   1. Repellents.
   2. Escape substances (inks of cephalopods, etc.).
   3. Suppressants (antibiotics, etc.).
   4. Venoms.
   5. Inductants (which cause galls and nodules).
   6. Counteractants.
   7. Attractants (attraction of the prey towards the predator).
B. *Kairomones:* adaptive advantage for the receiving organism.
   1. Chemical lures.
   2. Inductants which stimulate adaptation (for example the factor responsible for spine-development in rotifers).
   3. Signals which warn the receiving organism about danger or toxicity.
   4. Stimulants (growth factors).
C. *Depressants:* waste products, etc., which inhibit or poison the receiving organism without resulting in any adaptive advantage for the producer.

II. *Intraspecific effects*
A. *Autotoxins:* waste products toxic for the producer without any advantage for other species.
B. *Adaptation auto-inhibitors:* limiting population levels to their equilibrium position in the environment.
C. *Pheromones:*
   1. Sex pheromones 2. Social pheromones 3. Alarm and defence pheromones 4. Territory- and trail-marking pheromones, etc.

---

receiving them. The artificiality of such a classification can easily be perceived in this context because polluting chemicals could very well have been included in a separate section all to themselves since their effects are both inter- and intraspecific. Mercury, which is released by man in the biosphere, will at first poison plankton, followed by fish and crustacea, and eventually man himself. Mercury can also be returned to man by a more direct route such as through drinking water. It must be noted, however, that by different criteria, the distinction between depressors and autotoxins still remains valuable and the first definition ought to retain its position, after a modification which reads as follows: depressors – substances causing inhibition or poisoning of both the producer and the receiver. In such a context, mention of evolutionary disadvantages is euphemistic.

Florkin (1966) introduced the term co-actone which applies generally to any substance which is responsible for an interaction. Law and Regnier (1971) suggested the term 'semiochemical' for interactions involving the transmission of a message between organisms, whether inter- or intraspecifically. Information transfer by means of molecules is undoubtedly an aspect of cybernetics. It will be seen that every molecule corresponds to a real signal which is potentially available. Automatic behaviour in insects is linked to the reception of such signals. The understanding of these messages forms a very interesting part of chemical ecology; it enables a better understanding of relationships and can lead to elegant controls of behavioural systems. Based on such knowledge, methods for the preservation of natural equilibria can be developed and constitute an ideal complement to the fight against pollution. Nowadays, an efficient method to eliminate tsetse fly, the vector responsible for sleeping sickness transmission, is to spray insecticides over large areas. Could one not avoid the resulting pollution and the obvious large-scale destruction of the environment by a better understanding of biology and insect biochemistry?

Chemical ecology illustrates a multitude of causal relationships. Discoveries, in the course of the past few years, have been so great as to warrant a selective approach in the text which follows. This selection will be made on the basis of illustrating by way of examples the main types of observed activities.

Chemical ecology represents an important turning point in evolution. It is through it that man attempts to understand and to control and therefore become aware of the destructive effects of his development. Will this awareness be sufficient and will it provide enough power for preventive measures? Excesses render obvious the need for cures. However, if a retroactive mechanism is established too late, it may only slow down the process without avoiding catastrophes which will in the end prevail. The benefits of chemical ecology are that it allows nature to be seen in a new light; the future of man's interactions with his environment can be contemplated from its constructive aspects. Nevertheless, as observed by Eisner (1972), the predictive value of chemical ecology is small and its impact on the future of human society is still negligible. Too many details concerning interactions between living organisms remain unknown and the long-term fate of molecules which are involved in auto-intoxication phenomena is often a mystery. According to estimates by Commander Cousteau (1970), in the course of the past twenty years, the intensity of life in the seas of the world has dropped by at least 30 per cent. We should not just be alarmed but be prepared to intervene quickly. The purpose of chemical ecology is to understand; unfortunately the powers of intervention lie at a different level.

The various relationships which will be discussed in this book all have a common factor. With the exception of the destructive effects of man on his environment, the purpose of these relationships is to preserve the species. Conservation and development of the species are the underlying motives of these interactions.

In order to achieve these aims, it is necessary to establish structures. Those that exist form a large part of the chemistry of natural substances. Scheuer (1973) remarked that this chemistry is always surprising, both in terms of the scope of the molecular architecture which is to be found and in terms of the biological processes which are responsible for such structures. Nature still remains a gigantic catalogue of models and the knowledge of such models relies on the close collaboration of biologists and chemists. If the scope for predictions is still very low, many substances display biological activities which lead to spectacular applications (e.g. antibiotics).

Certain fields still have not been subjected to intensive investigations, such as the pharmacology of marine life. The scope of investigations is boundless and remains a constant source of astonishment and of interest, and we suspect than man's imagination will fail long before nature's resourcefulness is exhausted.

Chemical interactions in nature

In *The Magic Mountain,* Thomas Mann (1931) tells of the misfortunes of the traveller who might carelessly fall asleep in the shade of *Antiaris toxicaria.* The leaves of these large trees excrete fine droplets forming a poisonous spray. *Antiaris toxicaria* contains cardiac glycosides (Juslèn *et al.,* 1963). The manchineel tree (American Euphorbiaceae) is dangerous, not only because of its fruit but also because of a corrosive latex seeping from branches which causes burns that are slow to heal. West Indians always carefully avoid this tree as it is known to be the cause of many accidents.

Pliny the Elder (Soderquist, 1973) reports that the canopy of the walnut tree (*Juglans*) poisons the plants which it covers. It is now

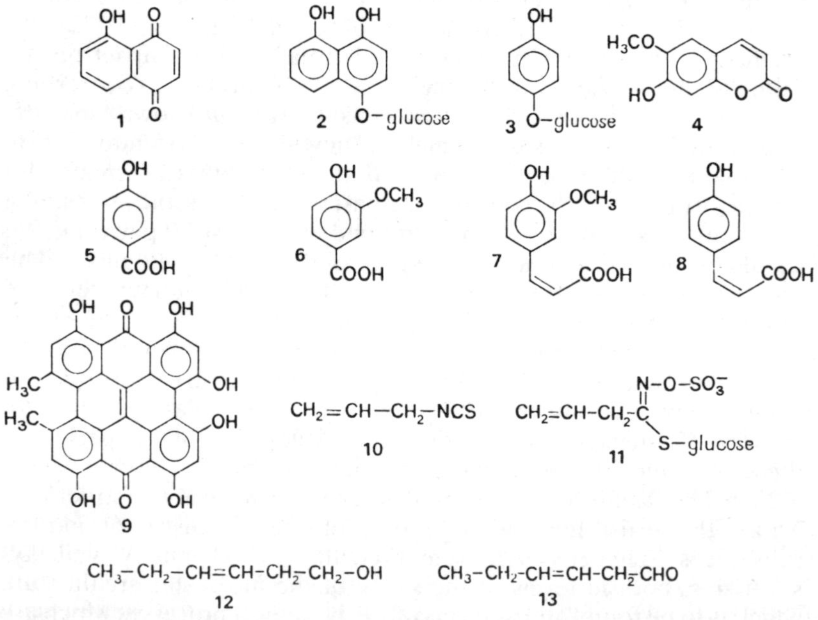

Fig. 1 Examples of chemicals responsible for allelochemical actions.

known that this phytotoxic effect is due to juglone **1** (5-hydroxy-1,4-naphthoquinone). This substance is found on the leaves and is washed to the ground by rainfall; juglone is toxic to the Gramineae, tomato plants, apple trees, etc. At a concentration as low as 10 p.p.m. tomato plant growth is reduced by 50 per cent and at a concentration of 100 p.p.m. lethal effects are 100 per cent (review by Soderquist, 1973). Furthermore, juglone is also toxic to small animals, on which it has a sedative effect, and has a marked effect on the growth of fungi and bacteria. High concentrations of this substance are found in the roots of walnut trees. It must be noted, however, that in the tree itself, juglone is not found as such but in the form of 5-glycosyl-1,4,5,-trihydroxy naphthalene **2** (Daglish, 1950); this glycoside is labile and releases trihydroxynaphthalene which is easily oxidized into juglone. This example of interspecific interaction is quite representative of a very widespread phenomenon in nature. Whittaker and Feeny (1971) suggested the use of the word allelopathy to characterize effects of that nature which are produced at a distance (allelochemical effect). In California, shrub communities (soft chaparral) which include *Salvia leucophylla* and *Artemisa californica* have a controlling effect on the Gramineae by the extensive production of volatile terpenes. These are thought to be taken up by the soil during dry periods by adsorption of the vapour. Bare areas, sometimes a few metres wide, surround these populations (Muller & del Moral, 1966). Similar effects have been observed in Mediterranean plant communities which include rosemary (Deleuil, 1950, 1951 a and b). Also in California, another type of plant community which is referred to as hard chaparral (*Adenostoma fasciculatum, Arctostaphylos glauca, A. glandulosa*) excretes phenolic compounds which, once dissolved by rainfall, inhibit the growth of grasses. The effect of fire is revealing: it is followed by vigorous growth of the Gramineae as long as the hard chaparral has not had time to reconstitute itself (see the superb review by Whittaker & Feeny, 1971). Among the identified chemicals, we find arbutin **3**, a hydroquinone glycoside which generates benzoquinone and *p*-hydroxy cinnamic acid (Whittaker, 1970a). *Encelia farinosa*, a plant found in desert areas of America, contains 3-acetyl-6-methoxy benzaldehyde, a growth inhibitor which is also drained into the soil by rain (Bonner, 1950; Gray and Bonner, 1948 a and b). Peach-tree roots contain large quantities of amygdalin which can release glucose, hydrogen cyanide and benzaldehyde. Cyanogenesis, the ability found in certain plants to release hydrogen cyanide, is a phenomenon which has been known for several centuries. Cattle and humans have often been victims of this phenomenon. Cassava-flour (Nigeria) can contain as much as 35 mg of HCN per daily intake, which is about half the lethal dose! In most cases hydrogen cyanide is produced from cyanhydrins deriving from sugars or it is obtained from amino-acid derivatives (including phenylalanine and tyrosine). About 800 plant species (70 to 80 families) are known to produce hydrogen cyanide (for review see Seigler, 1975).

A coumarin substance, scopoletin **4**, is stored in the roots of oats (Rademacher, 1941; Martin & Rademacher, 1960). Various substances displaying growth inhibiting effects have been found in the sunflower *Helianthus annus,* such as chlorogenic acid and scopoletin (Wilson & Rice, 1968). Heather produces chemicals which retard the growth of mycorrhiza and by this means repress the development of forests (experiments with fir and pine trees; Harley, 1952).

The seed germination of *Striga lutea,* a parasitic plant, is stimulated by strigol, a substance which is released by the roots of the host-plant (MacAlpine *et al.,* 1974).

The list of observed allelopathic effects which have been observed is already extensive. In many cases, growth inhibition is caused by normal plant constituents: terpenes, aromatic acids, phenols, etc. (*p*-hydroxy benzoic acid **5**, vanillic acid **6**, ferulic acid **7**, *p*-coumaric acid **8**, etc.). These phenols can be metabolized by soil bacteria and transformed into quinones; the reactivity of the latter with cell constituents (amine groups of proteins) is well established. The general rule is that these chemicals are not toxic for the producing plants; however, some cases of auto-intoxication have been reported where plant populations have been intoxicated by their own products (*Eucalyptus* for example; review by Whittaker & Feeny, 1971). The fact that in most cases these substances are not toxic for the producing organism can be tentatively explained by the existence of inactive glycoside precursors or by storage in vesicles on in barks.

From the point of view of evolution, the significance of chemical interactions between plant species is not easily grasped. If, in some cases, the adaptive advantage of one species over another is marked, more frequently the observed effects seem solely due to chance meetings; what appears to happen is that a metabolic end-product which is secreted, in a similar way to any waste product, has a detrimental activity on a particular plant of the neighbourhood and not on another, sometimes on the producing plant itself and sometimes not. The evolution of plant populations must rest on the complexity of chemical interactions which lead to equilibria, thus giving the impression of an inherent logic.

Some *Eucalyptus* trees inhibit the growth of other plants in the undergrowth (in the USA but not in Australia), thus demonstrating the adaptive powers of species (Whittaker, 1970b). The example of auto-intoxication of some tree species in tropical forests is of interest: the seeds can only sprout on areas already taken up by other species; the phenomenon of auto-intoxication which, from the point of view of species' evolution is negative, is in this instance positive since it leads to the extension of the area occupied by the species in question (Webb *et al.,* 1967). It is because these plant interactions are so complex that they are for the most part still unknown, particularly in the case of multiple associations. Deleuil (1954) describes the association of a composite, *Hyoseris scabra*, which grows in the presence of the bulb *Allium chamaemoly*, if a third species, *Bellis annua,* is also present. The role of chemical interactions in the formation of plant populations and the mechanisms by which various species are introduced still remains to be established and represents a wide-open field for investigation which would require the collaboration of ecologists, biologists and chemists.

The interaction between plants and animals makes use of a wide variety of means: plant defence mechanisms start with thorns and spines and extend to the synthesis of alkaloids, cardiac glycosides, steroids, etc. At a more elementary level, the plant might just be distasteful, bitter or difficult to digest (rich in tannin-protein combinations). Can one imagine more sophisticated levels of evolution than the synthesis by plants of insect hormones? Caterpillars of *Cecropia* will not survive the ingestion of food containing one part per thousand million of ponasterone A (Riddifort, cited by Whittaker & Feeny, 1971). Juvabione **127**, an analogue of juvenile hormones **123–125**, which has been isolated from certain woods (*Abies balsamea*) prevents the transformation of larvae into adults (see below). The biological activity of juvenoids is not very specific and starts with terpenes such as farnesol; however, the biosynthesis by plants of ecdysones is quite remarkable. It requires a series of chemical reactions which need considerable amounts of energy but, at the present, no biological significance for these substances is known. It is true, however, that the same conclusion can be reached for many of the substances which have been isolated from plants, alkaloids in particular.

Substances which mimic the effect of 1,25-dihydroxycholecalciferol and which provoke hypercalcification in herbivorous animals have been found in various plants: *Solanum malacoxylon* in Argentina and Brazil, *Trisetum flavescens* in Germany, *Cestrum diurnum* in the United States (Wasserman *et al.,* 1975).

Hypericin **9** found in *Hypericum* is accumulated by herbivores and causes skin photosensitivity leading to intense irritation (Brockmann *et al.*, 1951). On the other hand it must be noted that hypericin is a factor which initiates feeding in the beetle *Chrysolina brunsvicensis* (Rees, 1969). A phenomenon which is often encountered is that by which a substance will specifically attract one species while it is repellent for others. Cruciferae contain allyl isothiocyanate **10** which is stored in the form of a glycoside (sinigrin **11**) and is enzymatically released when the plant is damaged. This chemical is an irritant and a repellent for many species but specifically attracts Lepidoptera of the genus *Pieris* (Whittaker & Feeny, 1971). The insect–plant interactions will be followed in more depth at a later stage as they include very complex mechanisms. Two effects are superimposed: the attraction by the host-plant or its family and repulsion by other plants. This adaptation enables the species to avoid toxic traps; the same substances will be used by the larvae for food recognition and will attract females which are egg-laying and in this way the cycle will be completed.

The need to attract insects during the flowering periods but to repel them at other times is sometimes reflected in variations in the equilibrium between the synthesis of hexenol and hexenal **12** and **13**; the former is an attractant and the latter a repellent. Attraction by means of the fragrant substances of flowers which thus improves pollination is of course a classical example of interspecific chemical interaction.

Under most circumstances, bees choose their food (pollen or nectar) by carefully avoiding plants with little nutritive value (including conifers) as well as those which might present some danger. Unfavourable conditions, however, can lead them to frequent ranunculuses, horse-chestnut trees and silver limes which will have disastrous effects (Louveaux, 1965). The toxicity of silver lime flowers is responsible for the death of many insects which, consequently, can be found at the foot of these trees during the flowering season. Herbivores are often the victims of poisonous plants such as hemlocks and ranunculuses. In the West of the USA a herd perversion has been observed in which cattle forage for toxic leguminous plants and it is tempting to compare this phenomenon with human toxicomania (Louveaux, 1965). It is well known that goats chew tobacco with evident pleasure, although they are not sensitive to nicotine. They also appreciate rhododendrons but these have rather poisonous effects. Cattle will also feed on tobacco, and in this way become intoxicated if by accident they get into tobacco fields.

We can see that the phenomenon of chemical interaction between living organisms is a general one. This interaction obviously starts at the level of the biosphere. Went (1964) estimated that the amount of volatile organic matter released in the atmosphere is of the order of one thousand million tonnes per year. Arpino *et al.* (1972) attempted measurements of the quantity of wax found in suspension in the atmosphere surrounding conifer forests; 1980 m^3 of air were filtered giving 18 mg of products, which were, for the most part, hydrocarbons. The release of particulate matter by pine trees might be attributed to a needle-effect by which cuticle waxes are transformed into aerosols by electric fields. The bluish halo which is sometimes observed above forests could be caused by such a solid aerosol. The amount of organic matter released into the oceans and rivers as well as that produced by living organisms, particularly plankton, is also considerable (for these topics, see for instance Saliot & Barbier, 1974; Boutry & Barbier, 1974). Particulate or dissolved organic matter plays a definite nutritive role in the oceans and takes part in complex cycles. Some of these solubilized molecules are thought to play other roles as well. The concept of chemical telemediators has been suggested by Aubert *et al.* (1971). Since telemediators are the means by which relationships between members of the same species or indeed different species are sustained, these substances are of utmost relevance to chemical ecology.

The mineral–plant interaction, and by way of plants the animal–mineral interaction, holds a preponderant role in nature. Plants absorb and lay down a large number of elements which are to be found in their environment and which they do not always require (plutonium for instance). The essential elements are: C, H, O, N, P, S, Ca, Mg, K, Fe, Mn, Cu, Cl, B, Mo, Co, Si, Se, etc. The storing of absorbed elements varies according to the species or to the organs under consideration (Miller & Flemion, 1973). Their biological role has been reviewed by Nason and MacElroy, 1963. The relative concentration of halogens, Cl > F > Br > I is inverted in a marine environment for fluorine and bromine. Camellia can store fluorides to concentrations as high as 3 g per kg. Fluoroacetic acid, which is found in *Dichapetalem cymosum*, is responsible for cattle poisoning in South Africa. This acid is fairly widespread in many plants to which it confers toxicity (Miller & Flemion, 1973). Organic molecules which contain chlorine or bromine are quite frequent whether in lower or in higher plants, on land or in the seas. Sea water contains about 50 μg of iodine per kg and certain algae can accumulate it in remarkable quantities; *Laminaria digitata* contains iodine up to 1 per cent of its dry weight (Blinks, 1951). A variety of iodinated tyrosine derivatives are formed and might be responsible for biological activities which are observed in the course of algae-bath treatments (thalassotherapy).

The accumulation of selenium from the soil in certain plants (*Astragalus*) is also dangerous for cattle (Shrift, 1969). Of the 500 species in North America, 25 are known to have toxic effects because of selenium fixation. *Astragalus pectinatus* growing on soil containing 2–4 p.p.m. of selenium contains up to 4000 p.p.m. Under identical growth conditions, maize only contains 1–10 p.p.m. The main synthesized metabolite is Se-methyl-selenocysteine. Many plant species which store selenium and which need it for growth and development are characterized by a repulsive smell, the intensity of which is related to the quantity of selenium stored (Rosenfeld & Beath, 1964). Dimethyl selenide has been identified in the volatile substance.

The accumulation of silica is often linked to physical defence mechanisms. The formation of thorns and sharp edges, etc. can depend on its concentration in specialized structures. The existence of silica gels in cells of the epidermis allows regulation of water evaporation.

The systematic analysis of the mineral elements in plants provide a method for the localization of mineral beds. Thus, in the USSR, arsenopyrite deposits were discovered by assaying the iron content of grasses. In the same country this method has been used in the search for deposits of boron, nickel, cobalt, copper, chromium and molybdenum.

Chapter 2

Preservation of the species: toxins, venoms and means of deception

The use of poison, by injecting it, applying it or throwing it at a distance by means of a sophisticated apparatus, is by its very nature an aggressive act. With the exception of man, aggressive behaviour is, however, rare and the attack is usually simply a defence reaction. The weapon, if suitable, is used in the hunt for food. In cases where the toxic substances are diffused throughout the body, or in cases where these substances are synthesized by a specialized gland but cannot be excreted, the defensive aspect of the toxins is rarely apparent. Poisoning of the predator appears to be due simply to the accidental encounter of two incompatible metabolisms. The assertion that evolution progressively brought about these mechanisms is rather vitalistic. At most we may believe that among all possible solutions those which have survived are the ones best adapted to life.

Bitter and toxic plants are shelters for insects which have chosen them as food supplies. The association of warning colours with particular chemical protection is another aspect of defence mechanism which is difficult to analyse in terms of evolutionary history. The phenomena that we observe in contemporary species seem to be the result of numerous struggles and of a few absolute prerequisites.

Habermehl (1975) suggested the classification of toxic animals into three categories. Under the term 'actively venomous animals' he classified all animals displaying aggressive behaviour. If on the other hand the venom-secreting organ were only used for defence purposes the animals were termed 'passively venomous'. If toxicity were purely accidental, for instance due to chance encounters, Habermehl used the term 'poisonous animal'.

In nature a wide range of toxic substances are found and a wide variety of ways have evolved in which to use them. The diversity of naturally occurring toxins is illustrated by the examples described below.

We shall start with mycotoxins. They are relevant both to animals and to man. St Anthony's fire or ergotism, which is caused by intoxication by rye ergot (*Claviceps purpurea*), used to cause terror throughout the whole of Europe in past centuries. The discovery of aflatoxins revealed that mycotoxicoses might have even more subtle origins. Scientific investigation of these toxins has gradually led to the establishment of controls and legislation.

Phytotoxins are of world-wide importance; the increase in agricultural production, particularly the production of cereals, oil-yielding plants, citrus fruits and fodder, is essential in view of the incessant increase in world population. Knowledge of phytotoxins might provide a stepping stone in the fight against pathogenic fungi and may indeed lead to the selection of resistant seed strains.

There are few examples in nature of biologically active molecules possessing such a variety of structure and degree of power as that shown by antibiotics. Although the first observation of an antibiotic effect was made long ago, the isolation and identification of the first antibiotic was delayed, partly because of the problems in coordinating the efforts of biologists and of chemists.

Mycotoxins, phytotoxins and antibiotics

Mycotoxicoses have been known for a long time in man and domestic animals. Ergotism, caused by alkaloids in rye ergot, was responsible for genuine epidemics during certain periods. The use of flours containing these alkaloids (which include ergotamine and other substances derived from lysergic acid **14**; Gröger, 1972) has been responsible for serious conditions such as convulsions and limb gangrene. Contamination of food by fungi is a widespread phenomenon, but it is only relatively recently that its importance has been fully acknowledged. The discovery of aflatoxins in peanuts and their derivatives (butter and oil) and of their influence on the development of liver cancers in animals and man, has revived the interest in the study of mycotoxins. The carcinogenic activity of other substances, such as penicillic acid **15** (Ciegler *et al.*, 1971) which has been recently demonstrated, has also stimulated research in this field. Mycotoxicoses present a serious threat on a world-wide scale by affecting the food sources of man and animals. It has been estimated that more than 10 per cent of all food losses can be linked to fungal contamination (55 million tonnes of cereals are lost every year).

The ability of fungi to synthesize a wide variety of toxic chemicals was observed during studies on the development of the production of antibiotics. Sixteen aflatoxins are known at present, with a structure either close to that of aflatoxin B_1 **16** or that of aflatoxin G_1 **17** (Detroy *et al.*, 1971; Roberts, 1974). The study of these substances was started after an epidemic of turkey X disease in England in 1960. The control of the contamination of foodstuffs by aflatoxin became very strict and the fight against infection by *Aspergillus* and *Penicillium* became widespread. Aflatoxins were first isolated from *Aspergillus flavus* and *A. parasiticus*; they have since also been identified in *A. niger*, *A. glaucus*, *A. ostianus*, *A. ruber*, *A. wentii* . . . and also in *Penicillium citrinum*, *P. digitatum*, etc.

Lopez and Crawford (1967) investigated the presence of aflatoxins in peanuts that were being sold for human consumption in markets in Uganda; they found that 15 per cent of all samples contained more than 1 p.p.m. of aflatoxin B_1 and 2.5 per cent contained more than 10 p.p.m. Statistics show that in Uganda children consume an average 570 g of peanuts per week.

Aspergilli can develop on various grains, such as rice and soya beans, during storage. Aflatoxins have been detected in animal feeds and in dehydrated foods for human consumption. In tropical areas many different foods can be infected by these fungi. The list of mycotoxins has become considerably longer during the past few years. Ochratoxin A **18** (*Aspergillus ochraceus*) is responsible for liver necrosis in animals feeding on infected foods (cereals, peanuts). *A. ochraceus* also contains penicillic acid **15** and melleine **19**, which is probably an ochratoxin precursor (see Steyn, 1971, for a review of this field).

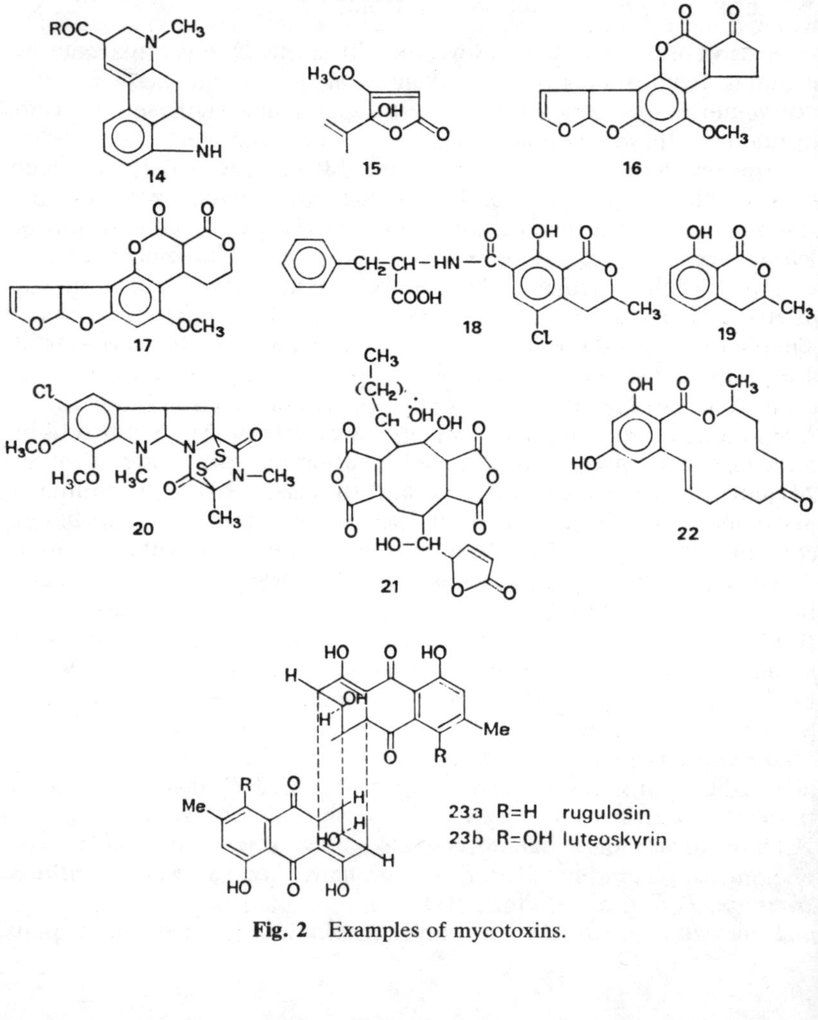

Fig. 2 Examples of mycotoxins.

Another group of mycotoxins which ought to be mentioned are the epipolythiadioxopiperazines, which include sporisdesmin G **20** (*Pithomyces chartarum*, a saprophyte of decaying Gramineae and a possible cause of facial eczema) (Taylor, 1971). *Penicillium rubrum* has been isolated from damaged wheat that had been known to cause cattle intoxication; rubratoxins, in particular rubratoxin B **21**, are responsible for this (Moss, 1971). The form of oestrogenism which affects pigs is caused by the presence of *Fusarium graminearum* (*Gibberella zeae*) in their food. The chemical which is responsible for this hormonal disturbance is zearalenone **22** (Mirocha *et al.*, 1971). Young males develop female sexual characteristics and females suffer from hyperfolliculinia symptoms, with a high abortion rate. Luteoskyrin **23 b** is part of a group of hydroxyquinones (rugulosin **23 a**, rubroskyrin) causing liver damage, which are synthesized by various *Penicillia* (*P. islandicum, P. rubrum, P. brunneum, P. tardum*) which usually grow on rice ('yellowed rice toxins') (Saito *et al.*, 1971).

In Moreau's book (1974) a detailed study of mycotoxicoses is presented. All the major fungi responsible for intoxications and the toxins which mediate the effects are listed, together with a description of the accompanying syndromes.

Chemical interaction between man and poisonous toadstools has been known for a long time. The Death Angel or Green Deathcap, *Amanita phalloides*, has suffered from a well-justified reputation in Europe for centuries. North America has its share of poisonous varieties, such as *Amanita virosa, A. verna, A. tenuifolia* and *A. bisporigera*; these are all as dangerous as the European species. *Amanita* poisoning is complicated by the delayed appearance of the symptoms, during which time serious liver and kidney damage may occur, leading to death. Our knowledge of *Amanita* toxins is due to the remarkable work of Theodor and Otto Wieland (1972). Two groups of toxic substances have been isolated: phallotoxins (such as phalloidin **24**) and amatoxins (such as α-amanitin **29**). The lethal dose for humans is 0.1 mg/kg for amanitins; a 50 g specimen of *Amanita phalloides*, containing about 7 mg of these toxins, is capable of killing a man. Amatoxins are bicyclic octapeptides which inhibit protein biosynthesis by acting on the DNA-polymerase of eukaryotic cells (which stops RNA transcription). Phallotoxins are heptapeptides. *Amanita phalloides* also contains an antitoxin, antamanide **35**, which if absorbed in large enough quantities and at the same time as the toxins, will protect the eater completely (5 mg/kg is required for mice). This observation, although remarkable, has limited the therapeutic use because the antitoxin has to be ingested before or at the same time as the toxin itself.

The cyclopeptides of *Amanita phalloides* are not the only substances responsible for causing death from eating toxic toadstools. Gyromitrin **36** (extracted from *Helvella* (*Gyromitra*) *esculenta, H. gigas, H. underwoodii*, etc.) is responsible for poisonings which display

Phallotoxins

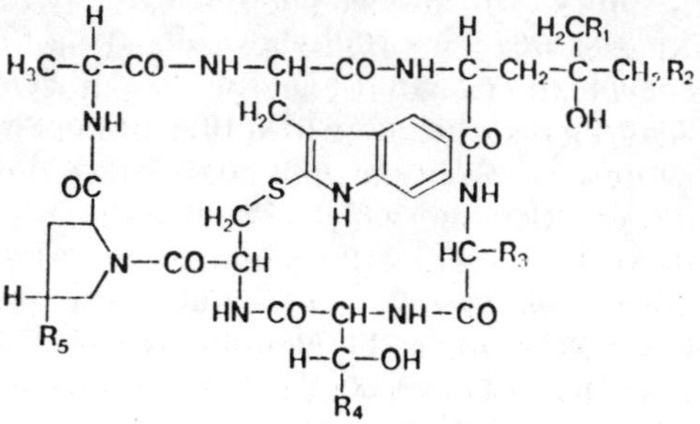

	R_1	R_2	R_3	R_4	R_5
24 phalloidin	OH	H	CH_3	CH_3	OH
25 phalloin	H	H	CH_3	CH_3	OH
26 phallisin	OH	OH	CH_3	CH_3	OH
27 phallacidin	OH	H	$CH(CH_3)_2$	CO_2H	OH
28 phallin B	H	H	$CH_2C_6H_5$	CH_3	H

Amatoxins

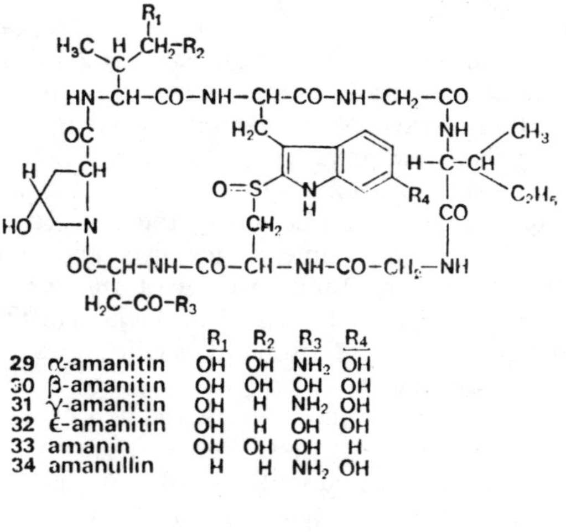

	R_1	R_2	R_3	R_4
29 α-amanitin	OH	OH	NH_2	OH
30 β-amanitin	OH	OH	OH	OH
31 γ-amanitin	OH	H	NH_2	OH
32 ε-amanitin	OH	H	OH	OH
33 amanin	OH	OH	OH	H
34 amanullin	H	H	NH_2	OH

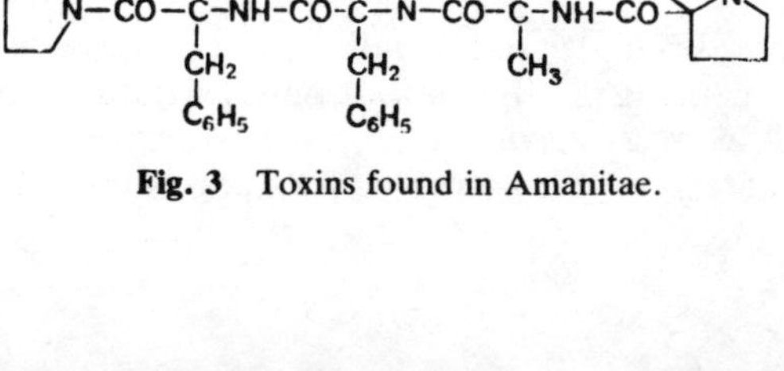

Fig. 3 Toxins found in Amanitae.

symptoms rather like those caused by *Amanita Phalloida.* Muscarin **37** (Wilkinson, 1961), which has been isolated from the fly agaric *Amanita muscaria* and *A. pantherina,* as well as from a number of species of *Inocybe* and *Clitocybe,* acts on the parasympathetic nervous system; 0.01 μg/kg is sufficient to lower the blood pressure of a cat and decrease the rate and amplitude of its heartbeat. Compounds which act as insecticides on flies and which, incidentally, are known for their narcotic and psychomimetic effects in man, have been obtained from *A. muscaria, A. pantherina* and *Tricholoma muscarium.* These compounds are isoxazoles **38** (muscinol), **39** (ibotenic acid), **40** (tricholomic acid) and **41** (muscazone). With the exception of ibotenic acid, these chemicals are responsible for damage to vision, mental confusion, loss of memory, spatial and temporal dislocation, and so on (Eugster *et al.,* 1965; Eugster & Takemoto, 1967; Benedict, 1972). Tricholomic acid is the most toxic, as far as the flies are concerned. The hallucinogenic powers of the small Mexican mushroom 'Teonacatl' (*Psilocybe*), also known as 'God's chair', were already known to Aztec

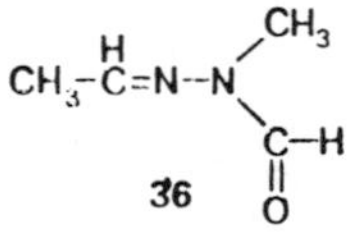
36

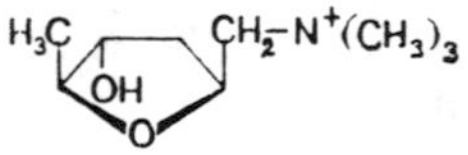
37

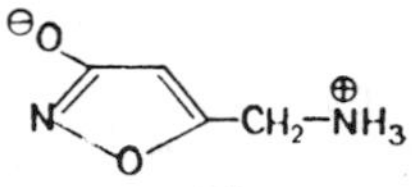
38

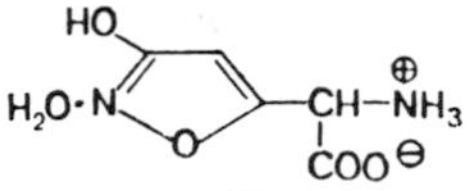
39

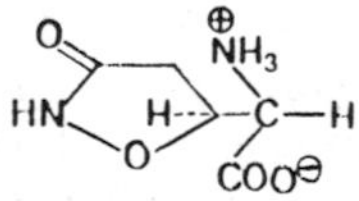
40

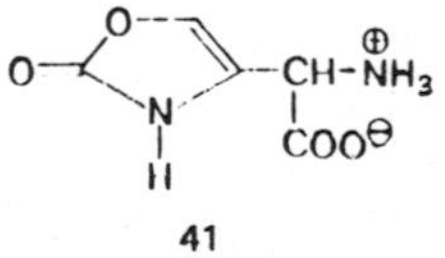
41

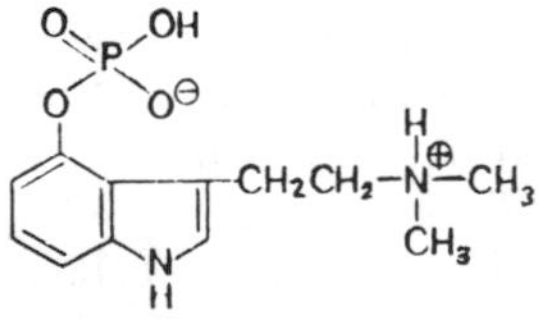
42

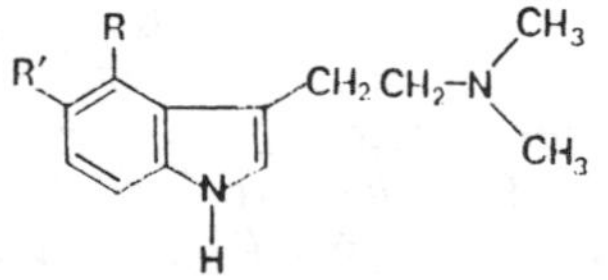

43 psilocin R′=H R=OH

44 bufotenin R′=OH R=H

Fig. 4 Fungal toxins.

priests. When the mycologist Watson became interested in this mushroom in 1958 it was still being used by the local Indians in witchcraft. Psilocybin **42**, psilocin **43** and bufotenin **44** cause hallucinations (Hofmann *et al.*, 1959; Benedict, 1972). In children, *Psilocybe* intoxication can cause long-lasting convulsions and death. Bufotenin, a constituent of toad venom, is also the main active ingredient in *Piptadenia peregrina* (Mimosaceae). Inhabitants of Peru and Colombia were using this plant for powder inhalations in order to achieve some sort of intoxicated state at the time of the discovery of these countries (Soleil & Lalloz. 1971). The entomopathogenic mushroom *Metarrhizium anisophae* contains a series of cyclodepsipeptides with insecticide properties which are characterized by several N——CH₃ groups (destruxins; Suzuki *et al.*, 1970).

The word mycotoxin is usually used to describe the toxins of mushrooms, fungi or toadstools which are responsible for illnesses in man and in animals, whereas phytotoxin is used for substances produced by mushrooms, fungi or bacteria which are phytopathogenic. Since food production can often be impaired by phytopathogenic micro-organisms, more knowledge of phytotoxins and the search for effective ways to limit their effects are of vital importance. A major problem at present is how to link the phytotoxicity of isolated substances studied by chemists with the observed plant diseases. This is aggravated by the fact that phytotoxins are isolated from cell cultures, and very rarely from the infected plant itself, with the resulting uncertainty as to the relevance of phytotoxins *in vivo*. Furthermore, the symptoms observed in any one infection are often due to several toxic effects being superimposed, which renders the comparison between the effect of a phytotoxin and that of a micro-organism even more difficult.

Lycomarasmin **45** and the aspergillomarasmins A and B (**46** and **47**) have been found in a species of *Fusarium* (Hardegger *et al.*, 1963; Haenni *et al.*, 1965; Barbier, 1972). These substances interfere with water metabolism in many plants, which accounts for the name marasmin suggested by Gaumann. They are derived from amino-acids which have become linked by bonds other than peptide bonds. Because of the position of two nitrogen atoms on two adjoining carbon atoms, the formation of complexes with ferric ions is possible. *In vitro*, the introduction of ferric ions increases the activity of marasmins, as can be assayed by the observation of tomato leaf withering.

The presence, in certain species of *Fusarium*, of all three phytotoxins **45**, **46** and **47** suggests that these might be genetically related. *Fusarium* spp. also produce fusaric acid **48**, which is responsible for the yellowing of infected plants. This might have its origins in cell permeability changes, resulting in large electrolyte losses. *Pseudomonas tabaci* infects tobacco plants. 'Wild fire toxin' is a dipeptide in which threonine is linked to a hydroxy-diamino-diacid **49** (Stewart, 1971). This chemical causes leaf chlorosis and is toxic to chloroplasts.

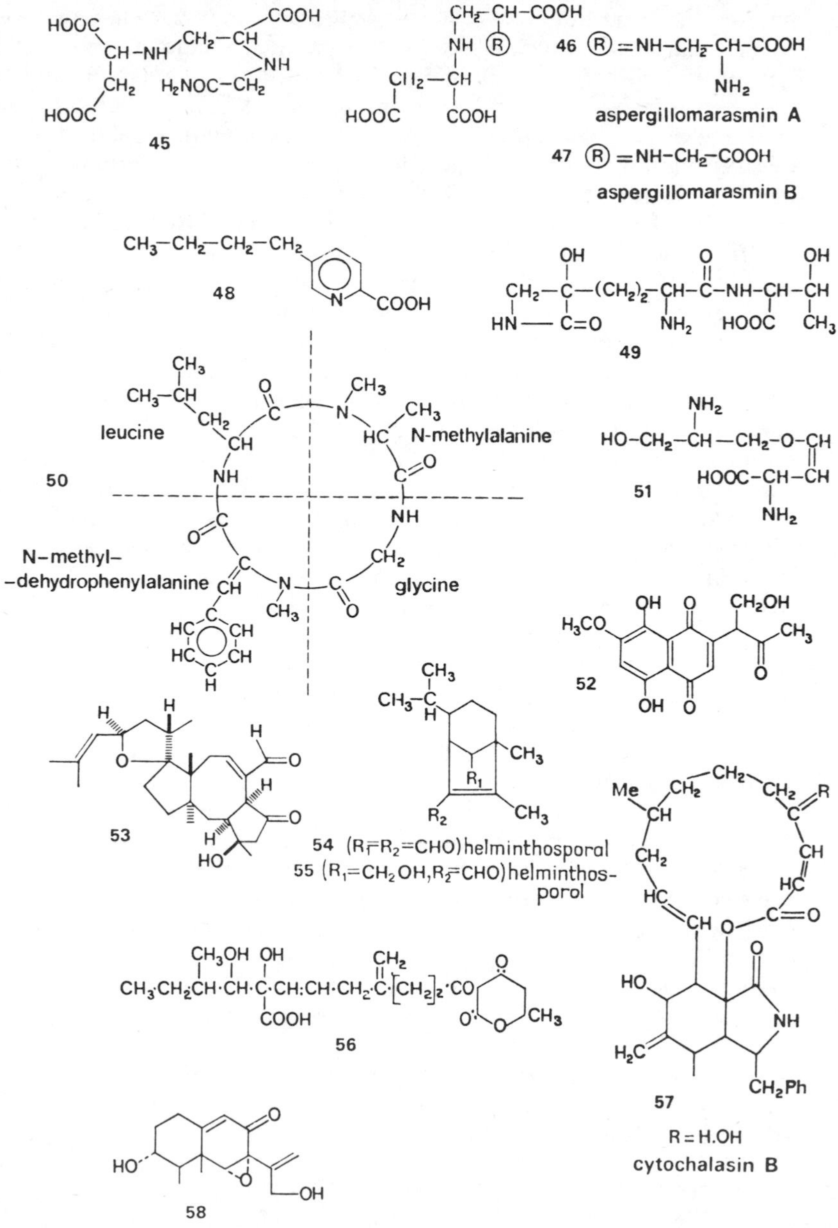

Fig. 5 Examples of phytotoxins.

Methionine and glutamine inhibit this process, which indicates that the chemical might act at the glutamine synthetase level. Tentoxin **50** (Templeton *et al.*, 1967), which has been isolated from several phytopathogenic *Alternaria,* is a cyclic tetrapeptide consisting of N-methylalanine, glycine, N-methyl dehydrophenylalanine and leucine. Soya leaf necrosis is due to *Rhizobium japonicum,* which synthesizes rhizobitoxin 51 (Owens *et al.,* 1972) that blocks the cleavage of cystathionine into homocysteine. Novarubin 52 is one of the four naphthazarins that have been isolated from *Fusarium solani* cultures (Kern, 1972). *Helminthosporium oryzae* infects rice leaves and in this instance the phytotoxin has been isolated from cell cultures *and* from the infected plant; it consists of ophiobolin A 53 (Canonica *et al.,* 1966; Wood *et al.,* 1972). Helminthosporal 54 has been obtained from cultures of *Cochliobolus sativus,* which infects cereals. This aldehyde acts by inhibiting phosphorylating oxidations in mito-chondria while the corresponding alcohol, helminthosporal 55 has gibberellinic action (de Mayo *et al.,* 1963; Casinovi, 1972). Alternaric acid 56 (Bartels-Keith, 1960) extracted from *Alternaria solani,* a pathogen of Solanaceae, has a withering effect on the leaves at a concentration as low as 5 µg/ml. Cytochalasins constitute a large group of toxins which have been isolated from *Phoma*-type fungi. Their structure is that of large cyclical lactones (e.g. cytochalasin B or phomine 57) which interfere with cytokinesis and cause nuclear extrusion. First isolated from *Helminthosporium dematioideum* and then from *Metarrhizium anisopliae* and *Rosellina necatrix* by Carter and by Aldridge *et al.,* their structure has been established by the parallel work of Aldridge *et al.* (1969, 1972), and that of Rothweiler and Tamm (1970); Carter (1972). The link between the phytotoxic activity of cytochalasin B and the pathogenicity of three stocks of *Phoma exigua* was demonstrated in 1972 (Bousquet & Barbier). *Phoma* synthesize many odd metabolites including phomenone 58, which inhibits seed germination (Riche *et al.,* 1974).

   The phytopathogenic activity of bacteria might well be of a physicochemical nature; polysaccharides causing leaf withering by obstructing sap flow are certainly known. The mode of action of certain more complex glycopeptides could be of a similar nature (Wood *et al.,* 1972; Lousberg & Salemink, 1972).

   These few examples illustrate the importance of chemical interactions for which mycotoxins and phytotoxins are responsible. The list of these chemicals is already long in spite of the fact that the field is young. Their biological mode of action is extremely varied and still ill-understood. Since they are active at very low concentrations, public health might easily be endangered and early detection of such toxins in food is therefore important. Since at the moment we ignore the biological significance of these toxins to the producing micro-organisms, they must be classified, according to Whittaker and Feeny's classification (1971), as waste products which poison the host. But except in the case of some mushrooms, which because of their toxicity are protected from man, it is difficult to see why these poisonous effects should be of any evolutionary advantage. Since in most cases the phytotoxins lead to the death of the host, an apparent disadvantage is a more obvious conclusion.

Some plants are able to synthesize antifungal substances when invaded by fungi. This is the case with potato tubers, which synthesize lubimin, hydroxylubimin (Katsui *et al.*, 1974) or phytuberin (Hughes & Coxon, 1974).

Vuillemin (1889) coined the word 'antibiose' or antibiosis in order to describe the toxic interactions between living organisms. In 1887 Pasteur and Joubert had noticed that in certain culture media, growth was inhibited by the introduction of aerobic bacteria. In 1929, Fleming observed the effect of *Penicillium* on bacteria but it was not until Waksman in 1940 had isolated penicillins that the word antibiotic became generally accepted. Originally, these substances were chemicals synthesized by micro-organisms, which inhibited development of another micro-organism at remarkably low concentrations. If we contrive to use this definition, antibiotics are a remarkable example of allelochemical interactions (see Table 1), where the evolutionary advantage lies obviously with the producer. Antibiotics belong to the most varied chemical families. A wide selection, together with a rich bibliography, can be found in the book of Asselineau and Zalta (1973) where interesting pharmacological notes can also be found.

The biological mode of action of antibiotics is still rather poorly understood. They often act as protein synthesis inhibitors by reacting with ribosomal structures. The resistance phenomenon can involve genetic modifications or the adaptation of enzymatic systems (for instance the hydrolysis of the β − lactam ring in penicillins). Other mechanisms, such as a reduction in cell permeability, might also be involved. The micro-organisms responsible for antibiotic synthesis belong to many families but are mainly lower fungi, bacteria and actinomycetales (*Streptomyces*). According to a study of Woodruff and MacDaniel (1958), out of 10,000 actinomycete strains isolated from the ground, 2500 synthesize antibiotics. Antibiosis seems to be a logical consequence of the never-ending warfare which is found in the microbial world.

Some characteristic terpene antibiotics are roridin A **59** (Bohner & Tamm, 1966) (*Myrothecium roridium*) and fusidic acid **60** (*Fusidium coccideum*). Fusidic acid is a protein synthesis inhibitor; it is not very toxic but highly active against staphylococci. Amino-glycosides and monosaccharides represent a particularly common group of antibiotics. Streptomycin **61** was isolated from *Streptomyces griseus* by Waksman *et al*. This micro-organism is usually inhibited from developing by streptomycin, which can therefore be considered as a toxic waste product and a growth-inhibiting factor. The industrial production of this antibiotic relies on the use of resistant strains. The action spectrum of streptomycin is wide; it acts on spirochaetes, mycobacteria, Gram-positive and Gram-negative bacteria and many

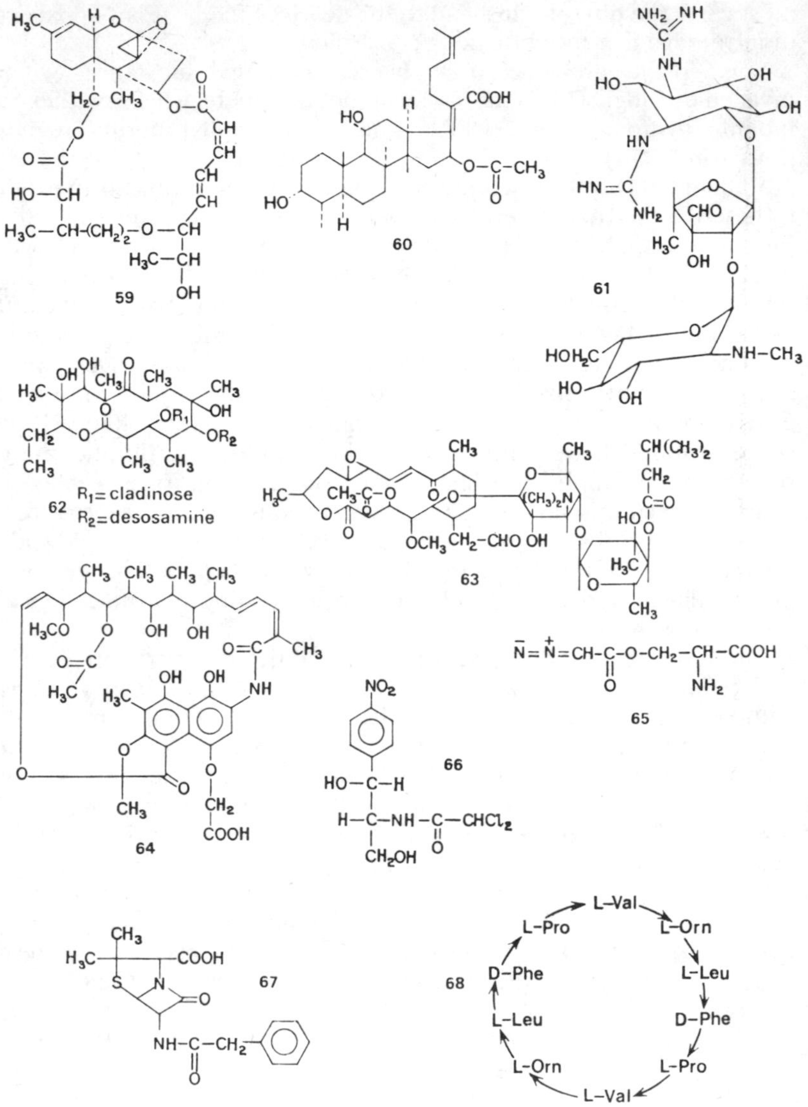

Fig. 6 Examples of antibiotics.

others. Streptomycin interferes with several bacterial metabolic pathways and affects protein biosynthesis by binding to ribosomal subunits and interfering with the codon–anticodon recognition.

Erythromycin A **62** (*S. erythreus*) is a macrolide (containing macrocyclic lactone) (Perun & Egan, 1969) which is used in combating staphylococci and pneumococci. It also acts by interfering with protein synthesis by binding to ribosomal structures. Other antibiotics in this

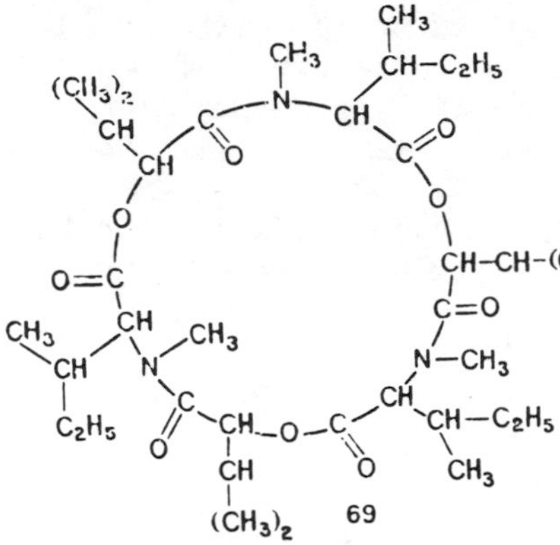

69

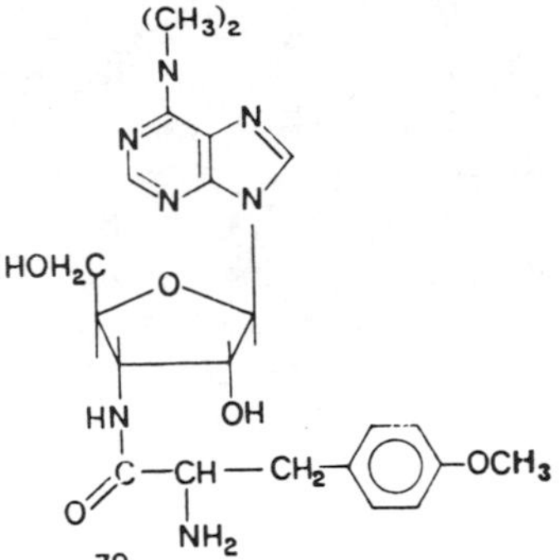

70

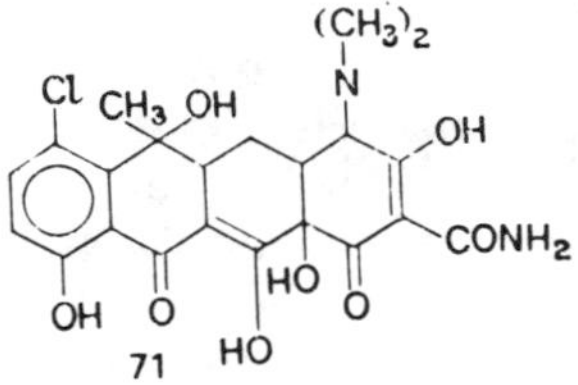

71

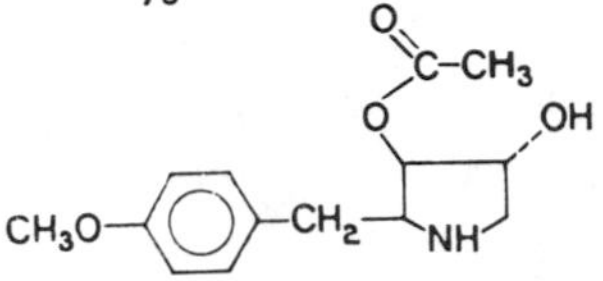

72

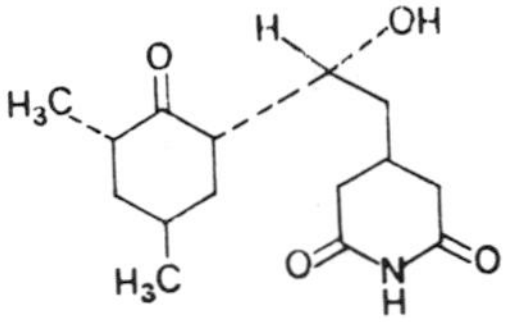

73

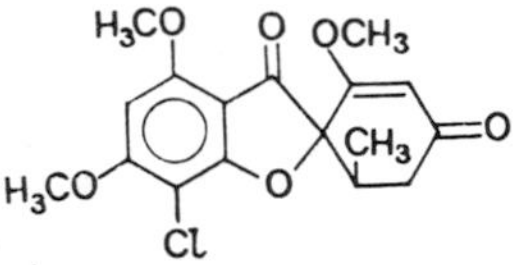

74

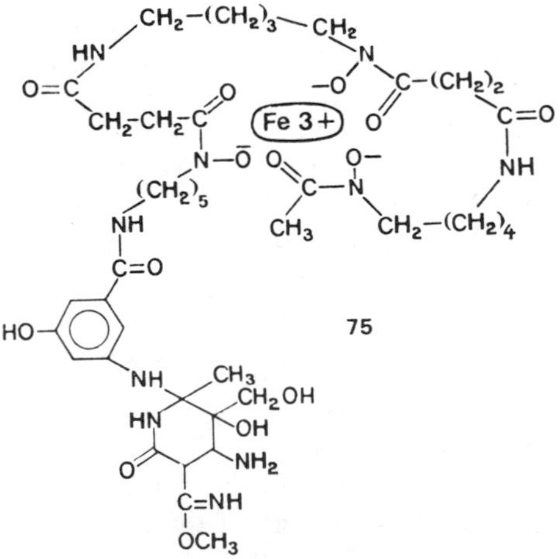

75

Fig. 6 Examples of antibiotics (continued)

group include magnamycin **63** (*S. halstedii*), nystatin (*S. noursei*; powerful fungicide) and rifamycin B **64** (*Streptomyces mediter-*

ranei) (Oppelzer *et al.*, 1964). The chemical structure of these substances is, however, not always so complex and sometimes surprisingly simple structures such as those of azaserin **65** or chloramphenicol **66** (the chloromycetin of *S. venezuelae*) are encountered.

The discovery of penicillins is still considered fundamental not only from the historical point of view, but also because of their medical consequences. After the classical work of Fleming, Raistrick attempted unsuccessfully to isolate these substances. Enriched preparations were obtained in 1938 by Florey and Chain (3% of penicillin). From 1942 onwards, 22 research groups in the USA and 17 in Great Britain cooperated to achieve the isolation and the characterization of the antibiotic. Since *P. chrysogenum* gave better yields, it replaced *P. notatum*. Penicillins (penicillin G **67**) are produced by many *Penicillium* species and even by some *Aspergillus*, for instance *A. flavus*. The synthesis of penicillin G (a valuable therapeutic antibiotic) can be facilitated by adding phenylacetic acid to the culture medium. It is a well-known fact that penicillins are active on Gram-positive bacteria, staphylococci and gonococci. Biological activity is due to the inhibition of peptidoglycan biosynthesis which forms cell membranes.

The *Bacillus* genus is characterized by the many polypeptidic antibiotics that it synthesizes. They contain D-amino acids (Bodansky & Perlman, 1969). For instance gramicidins, such as gramicidin S **68** (*Bacillus brevis*) act on Gram-positive bacteria (staphylococci, streptococci). Enniatin A **69** (*Fusarium orthoceras*) is a depsipeptide, which is cyclized by the formation of peptide- and ester-linkages (presence of hydroxy-amino acids; Russell, 1966).

Nucleosidic antibiotics differ by their interesting property of acting not only on bacteria, but also on certain protozoa (trypanosomes) and on some cancers (puromycin **70**; *S. alboniger*).

Tetracyclines which have polycyclic structures are known for their wide activity spectrum which includes rickettsiae and viruses. The elucidation of their structure is essentially due to the work of Woodward *et al.* (review by Clive, 1968). One of the best known tetracyclines is aureomycin **71** (from *Streptomyces aureofaciens*).

Anisomycin **72** which is obtained from a strain of *Streptomyces* is active in protozoa. Cycloheximide **73** (*S. griseus*) is a glutaric imide derivative and is too powerful to be used clinically but, since its activity spectrum includes many fungal species, it is useful for agricultural purposes. Cycloheximide is the most repulsive substance known to rodents; rats have chosen to die rather than to drink water containing as little as 5 p.p.m. of the chemical!

The low toxicity of griseofulvin **74** (*Penicillium griseofulvum*) makes it a useful drug for use against mycoses.

Some antibiotics contain a chelated metal ion (iron or copper) such as sideramin and sideromycins. The first of these, grisein, was isolated in 1947 from *S. griseus* by Waksman. These compounds are powerful and have a wide spectrum, but resistance to them occurs readily. Ferrimycin A_1 75 is said to be 50 times more active than penicillin.

Toxins in marine invertebrates

Unlike plants, animals cannot allow part of themselves to be destroyed in order to poison their enemies. Their action must protect their integrity, must be as fast as possible and, if unable to kill their opponents, must at least help repel the attack. Since the efficiency of a chemical weapon relies in part on its originality, a wide variety of toxic chemicals have been elaborated. Florkin (1966) suggested the general term 'co-actone' to define any chemical messenger between living beings (whether inter- or intraspecific). Defence substances give their producers the evolutionary edge and are, according to the classification of Brown *et al.* (1970) 'allomones', among other things. Most of the identified toxins have only recently been discovered and their structure sometimes reveals unexpected features. The number of known substances is still quite small compared with that of all the sea animals believed to be toxic or venomous.

Marine annelids contain nereistoxin 76, a tertiary amine with a cyclic disulphide (analogy with lipoic acid; Okaishi and Hashimoto, 1962 a and b). These animals are able to inject poisons either through bites or by their bristles (review by Halstead, 1965). Nereistoxin and its synthetic derivatives have insecticidal characteristics (Baslow, 1969). *Paranemertes peregrina* stores in its proboscis a paralysing substance, anabascin 77 (Kem *et al.*, 1971).

Various choline esters have been isolated from the hypobranchial glands of marine gastropods: murexine 78 from *Murex* (Erspamer & Benati, 1953), seneciolylcholine 79 from *Thais floridiana* (Whittaker, 1959) and acrylylcholine 80 from *Buccinum undatum* (Whittaker, 1960). Injection of murexine into mice causes muscular paralysis and respiratory collapse. Cone shell venoms, in particular that of *Conus geographus,* can be fatal for man (Rice & Halstead, 1968). These venoms contain a mixture of amines, peptides and proteins which are biologically active and also N-methylpyridinium, homarin 81 and γ-butyrobetain, which have curare-like effects. The saliva and salivary glands of whelks also contain tetramethyl ammonium which has a curarizing effect on vertebrates. Structure 82, which has been determined by X-ray crystallography (Kosuge *et al.,* 1972), is surugatoxin, which is found in the carnivorous marine gastropod *Babylonia japonica*. Its structure is unusual in that it consists of bromo-indole bound to a pteridine derivative. Consumers of this shellfish may suffer from visual and speech disturbances. Species of *Aplysia* are able to concentrate bromo-terpenes which they obtain from the algae they feed on. Aplysin 83 and aplysinol 84 have been isolated from *Aplysia kurodai* (Matsuda *et al.*, 1967); they have both been obtained by Irie *et al.* (1969) from algae of the *Laurentia* family. If aplysin is added to the feed of mice, it causes hypersalivation, ataxia and death by respiratory paralysis. Aplysiatoxin (Kato & Scheuer, 1974) is a macrocyclic dilactone linked to a bromophenol. Sea-cucumbers (Holothurians) secrete, by means of specialized organs called Cuvier glands or Cuvierian organs, certain sulphated steroid glycosides which are extremely toxic and are called holothurins. Holothurin A which is produced by *Actinopyga agassisi* (Chanley *et al.*, 1960) contains four sugars: D-xylose, D-glucose, 3-O-methyl-D-glucose and D-quinovose. The structure of the aglycone obtained by hydrolysis, 22,25-oxydo-holothurinogenin **85**, is a lanosterol derivative (Scheuer, 1971; Habermehl & Volkwein, 1971). The 7,9-diene is an artefact obtained during hydrolysis. An experiment in favour of this hypothesis has been carried out by Chanley and Rossi (1969) (Grossert, 1972) in which careful elimination of sulphate was carried out. A neo-holothurinogenin is liberated with a double bond in position 9 but not in position 7 and a hydroxyl residue in position 12. Aglycones with closely similar structures have been described by various authors studying a variety of species (Tursch *et al.*, 1970; Roller *et al.*, 1969; Chanley *et al.*, 1966; Scheuer, 1971). Holothurin A inhibits the transmission of nerve impulses in rats and frogs (for a review of physiological effects, see Baslow, 1969). Hashimoto and Yasumoto (1960) have reported the existence of toxic saponins in starfish. This observation emphasises the close phylogenic relationship between starfish and holothurians, since ophiuroideans and crinoideans do not contain and of these substances. This relationship had been presumed because of the sterol distribution and that of naphthoquinone pigments in echinoderms (Gupta & Scheuer, 1968; Singh *et al.*, 1967). Using *Asterias amurensis* as starting material, Yasumoto and Hashimoto extracted a mixture of six glycosides, the predominant components being the asterosaponins A and B. Hydrolysis yielded D-quinovose, D-fucose, D-galactose, D-xylose and sulphuric acid. The main aglycone was pregnene-diolone **86** (Ikegami *et al.*, 1972 a; Sheikh *et al.*, 1972). Two other genins, with a structure deriving from cholesterol (for instance 3,6,23-trihydroxy-5α-cholest-9(11)-ene) have been published by Turner *et al.* (1971) and Ikegami *et al.* (1972 b). These saponins are thought to inhibit the spawning-inducer hormone in starfish (Ikegami *et al.*, 1972 b).

Little is known about the composition of coelenterate venom (Halstead, 1971). The biological properties of the toxins secreted by nematocysts have been reviewed by Humm and Lane (1974). The venom of nematocysts of cnidaria, sea-anemones and jelly-fish consists essentially of proteins (Larsen & Lane, 1966; Crone & Keen, 1971; Ferlan & Lebez, 1972). Tetramethylammonium is the predominant active ingredient in the tentacles of the sea-anemones *Anemonia sulcata* and *Actinia equina* (2 mg/g; Mathias *et al.*, 1960). Octopuses also secrete toxic proteins and glycoproteins from their salivary glands (Ghiretti, 1959; Erspamer & Anastasi, 1962; Sacrin & Boissonnas, 1962). Moore and Scheuer (1971) obtained a toxin, palytoxin, from *Palythoa* (Zoantharia, Coelenterata) which is the most potent toxin known (LD$_{50}$ 0.15 μg/kg in mice), and this substance, which has a molecular weight of approximately 3300, is still being studied. In Hawaii, the collecting of *Palythoa* is prohibited by a local taboo; Scheuer's laboratory was destroyed by fire on the same afternoon that he spent collecting these animals, which illustrates for the first time a form of interaction not predicted by the table of Whittaker and Feeny (1971)!

Dinoflagellates of the type *Gonyaulax, Gymodinium, Prorocentrum* and *Prymnesium* can sometimes transmit a toxin to marine

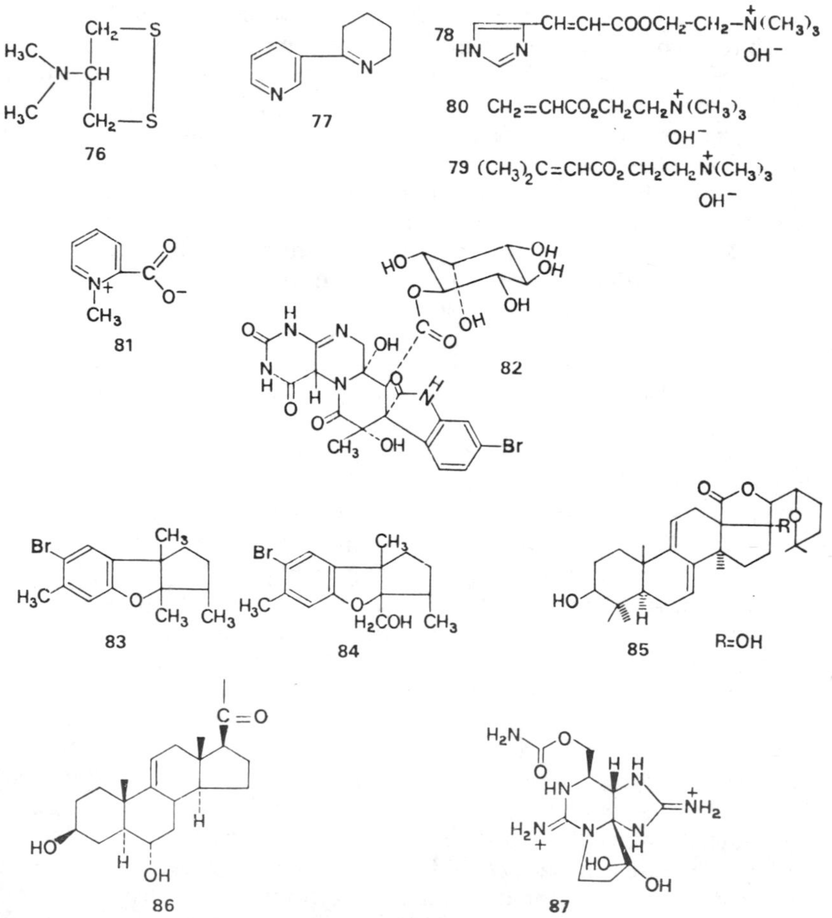

Fig. 7   Some of the toxins found in marine invertebrates.

invertebrates which feed on them, a fact which should not be overlooked by man, especially when these invertebrates happen to be oysters, clams or mussels! Saxitoxin is responsible for the toxicity of the Alaskan clam *Saxidomus giganteus* and of mussels and is identical to the paralysing toxin synthesized by the dinoflagellate *Gonyaulax catenella* (Wong *et al.*, 1971; Schantz *et al.*, 1957, 1966, 1975). This neurotoxin is also one of the most powerful toxins known (LD_{50} 5–10 μg/kg for mice). Those wishing to read reviews of the toxicity of marine invertebrates are recommended to consult Phisalix, 1922; McMichael, 1971; Premuzic, 1971; Scheuer, 1973.

Chemical defence in arthropods

In the case of arthropods, evolution led to a wide range of techniques both in terms of the poison used for defence (or possibly for attack) and in terms of the method used to reach the opponent. The enemy is not always killed by the toxic molecule but the repelling effect is often enough to save the arthropod's life. Vertebrates, including man, can be sensitive to these substances; an encounter between an epidermis and the sting of a bee or scorpion is not one to be recommended. According to the classification summarized earlier in Table 1, the defence chemicals of arthropods are allomones, namely substances which confer an evolutionary advantage upon the producing organism. Stinging hymenopterans such as bees, wasps and hornets have sophisticated systems enabling them to inject venom. The venom glands of these three species contain histamine, phospholipase A and hyaluronidase. Wasps and hornets in addition have phospholipase B and serotonin. Acetylcholine is found in hornets' venom. The venom glands also contain small proteins and peptides that can be very active such as apamin and mellitin in the bee. Apamin (18 amino-acids) exerts its effect on the central nervous system and on the vascular permeability of skin (LD_{50} 4 mg/kg in mice). Mellitin (26 amino-acids) is the main component of bee venom; it has haemolytic properties, causes contraction in muscle and is a neurotoxin acting on ganglionic synapses, with an LD_{50} of 3.5 mg/kg in mice (Habermann, 1971).

Scorpion venoms contain a few very active proteins: eight in the case of *Centruroides*, six or seven in *Tityium* and six in *Buthus judaicus*. The work of Miranda and Lissitzky (1958, 1961) and of Rochat *et al.* (1972) concerned the venom proteins of the North African scorpions *Androctonus australis* and *Buthus occitanus*. In 1967, Rochat *et al.* showed that a common sequence exists in scorpion neurotoxins. These authors isolated four basic neurotoxins with an average molecular weight of 7000 (Miranda *et al.*, 1970) consisting of a single polypeptide chain with four disulphide bridges. The complete sequence of these proteins was reported (LD_{50} 10 and 20 μg/kg in mice). 5-Hydroxy tryptamine which is also found in these venoms does not seem to be involved in the actual toxicity but is thought to be responsible for the intense local pain (Balozet, 1971). The risk of death for man following a sting by *A. australis* is of the order of 2 per cent and rises to 8 per cent in the case of children. With *Leiurus quinquestriatus*, the death-rate in children is 50 per cent and the Durango scorpion (*Centruroides*) in Mexico caused 1600 deaths between 1890 and 1926 in this single town of 40,000 inhabitants (Bucherl, 1971)! Spider bites are also responsible for many accidents. In humans attacked by *Phoneutria nigriventer* (South America), in addition to an unbearable local pain, visual damage, prostration and eventually death are observed. The venom consists of histamine, serotonin, polypeptides and a proteolytic enzyme (Schenberg & Pereira-Lima, 1971). Latrodectism disease in Mediterranean areas affects man as well as animals. In the former, a delayed pain appears which becomes unbearable and is accompanied by cramps, various disorders, facial oedema and conjunctivitis, anuria, tachycardia, muscular rigidity, etc. The death-rate is low. The toxin effects may last for as long as eight days, leaving the victim in a state of psychasthenia (Maretic, 1971). Various *Latrodectus* species can be responsible for these symptoms, in particular *L. mactans tredecim-gutatus*. The cephalothoracic glands of South American spiders of the family Gonyleptidae secrete a mixture of volatile quinones, known as gonyleptidine, which has an antibiotic activity on 18 species of bacteria and protozoa. According to Estable *et al.* (1955) and Fieser and Ardao (1956), gonyleptidine is a mixture of 2,3-dimethyl, 2,5,-dimethyl and 2,3,5-trimethyl benzoquinones **88, 89** and **90**.

An arachnid found in tropical areas, *Mastigoproctus giganteus* (*Pedipalpida*) which is 2 to 5 cm long, can eject liquids to distances of 20 to 80 cm. Analysis reveals that these consist of 84 per cent acetic acid, 5 per cent caprylic acid and 11 per cent water. Caprylic acid is thought to facilitate penetration by acting as a wetting agent (Eisner *et al.*, 1961).

Scolopendra bites are reputedly painful but they rarely cause death in man (LD$_{50}$ in mice, 1/300th of a *Scolopendra viricanis* gland). The composition of these venoms is not well known (Phisalix, 1922; Bucherl, 1971). When millipedes are threatened they release from a series of glands found on each segment, defence secretions containing a large amount of *p*-benzoquinones (Behal & Phisalix, 1900). Benzoquinone **91**, toluquinone **92** and 2-methoxy 3-methyl benzoquinone **93** have been identified in various species (Barbier & Lederer, 1957; Barbier, 1959; Schildknecht & Weiss, 1961; Eisner *et al.*, 1961; Wheeler *et al.*, 1964; Eisner & Meinwald, 1966; Weatherston, 1967; Périssé & Sales, 1970). It is likely that these substances protect the animals from their predators, in particular ants.

In humans, these substances cause skin irritations, burns and rashes, which are often found on the skin of those falling asleep outdoors. Quinones which can cause blistering and which easily bind to protein amine groups are thought to be responsible for these effects.

Rhinocrichus lethifer, a species found in Haiti, can throw its venom distances of 50 to 100 cm as a fine spray. Desquamations and phlyctaena sometimes appear as a result. *Rhinocrichus* often blinds small animals which eventually die of starvation since they are unable to feed. Surprisingly, some of the exotic species are used in the preparation of arrow poison (Burtt, 1947).

The millipedes *Apheloria corrugata* and *Pseudopolydesmus serratus* release hydrogen cyanide when disturbed (up to 645 μg for the first species, or many times the amount required to kill a mouse) (Eisner *et al.,* 1967). These millipedes have a sophisticated apparatus for the production of hydrogen cyanide, consisting of a two-chamber reactor (Fig. 8). One of these chambers is a mandelonitrile **95** reservoir (a combination of benzoic aldehyde and hydrogen cyanide). A muscle controls a valve leading to the second chamber where mandelonitrile is enzymatically dissociated (Eisner *et al.,* 1963 a and b). Secretion of the toxic substance can last for more than half an hour and provides an excellent defence mechanism since the gas also has a repulsive effect, on ants for instance.

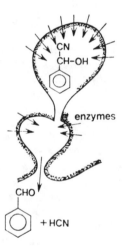

Fig. 8 'Reaction chamber' of the millipede *Apheloria.*

The existence of the cyanhydrin of a cuminaldehyde glycoside has been reported in the millipede *Polydesmus (Fontaria) vicinus*: this molecule could be responsible for hydrogen cyanide production in this species (Pallares, 1946).

Schildknecht *et al.* (1966) and Meinwald *et al.* (1966) identified two quinazolinones **96** and **97** in *Glomeris marginata*. When attacked, this myriapod rolls up in a ball and releases from between the segments of its body droplets of a convulsive toxin and a glue which solidifies in air. These quinazolinones are insidious poisons. If, say, a spider attacks *Glomeris*, it easily wins the fight, but a few hours later, may become paralysed. This intoxication is caused by glomerine **96** and homoglomerine **97** (Schildknecht, 1971).

Certain caterpillars have stinging bristles which can cause cutaneous reactions in man and animals (lepidopterism). In Europe, the pine-tree caterpillar (*Thaumetopoea pityocampa*) is well known for such misdeeds. The list of species causing such urticaria is quite long (Picarelli & Valle, 1971; Pesce & Delgado, 1971; Rotberg, 1971). Butterflies of the *Hylesia* genus (South America) have their bodies covered with hairs resembling poisoned darts which can cause dermatitis and allergies. In areas inhabited by these insects, accidents occur either by direct or indirect contacts since the hairs are easily detached when the animal is flying. In places where many of these butterflies are found, real hair clouds have been seen. Phisalix (1922) had already noticed that the moths of the *Zygaena* genus can kill mice by injection. Histamine and hydrogen cyanide have been shown to occur in *Zygaena lenicerae* (Bisset *et al.*, 1960; Jones *et al.*, 1962). *Callimorpha jacobea* contains large quantities of histamine (750 μg/g) and at the same time can accumulate the alkaloids of *Senecio* (Bisset *et al.*, 1960; Aplin *et al.*, 1968). Histamine has also been found in some Arctiidae.

Von Euw *et al.* (1969) demonstrated the presence of aristolochic acid **104** in the butterfly *Pachlioptera aristolochiae*. This toxic substance comes from plants the larvae live on. Caterpillars of the Notodontidae family excrete formic acid for defence purposes (Schildknecht and Schmidt, 1963). In addition to this acid *Heterocampa manteo* also utilizes 2-undecanone and 3-tridecanone (Eisner *et al.*, 1972). Formic acid, which is used efficiently as a defence mechanism by many caterpillars, had already been identified in these animals by Poulton in 1888. Caterpillars of the Papilionidae can unfold an organ which appears like a double horn and which produces volatile substances (osmeterium). The active mixture in *Papilio machaon* is of isobutyric and 2-methyl butyric acids (Eisner & Meinwald, 1965), which provide effective defence against ants. This procedure is widespread among many Papilionidae species (Eisner *et al.*, 1970). The larvae of the danaids *Danaus plexippus* and *D. chrysippus* which feed on *Ascelepias* (milkweed) can concentrate toxic cardiac glycosides which have been studied by Reichstein *et al.* (1968) (calactin and calotropine).

These poisons not only render the caterpillars and butterflies bitter, indigestible and toxic to vertebrates (mainly birds) that attempt to feed on them but they also protect the larvae which live on *Asclepias* since the latter are protected against herbivorous animals for the same reasons.

Toxic cardiac glycosides have also been observed in Orthoptera *Poekilocerus* and *Phymatus* (von Euw *et al.*, 1967; Reichstein, 1967). The injection of venoms by ants causes painful reactions but is rarely fatal in man. Nevertheless these insects are pests in areas where they are found in large numbers, such as in the Amazon. Ants use their venom not only for defence purposes but also to kill the small animals, mainly insects, that they live on. The venom gland of *Formica rufa* is essentially filled with formic acid (20 to 70%) (Weckering, 1960). In *Myrmicaria natalensis,* acetic acid predominates (35%), followed by isovaleric acid (31%) and propionic acid (22%). Ant secretions can have multiple roles such as defence, food, territory marking, alarm, etc. These have been discussed by Bergström and Löfqvist (1973). These authors examined the contents of the Dufour glands in three *Formica* species. This gland releases it contents at the same time as the release of formic acid from the venom gland takes place in the course of the defence reaction. Forty-six compounds have been identified, including hydrocarbons, geranylgeraniol, octadecylacetate, etc. The anal glands of the Argentian species *Iridormyrmex humilis* secrete a substance with insecticide properties (Weckering, 1960; Pavan, 1959). Among such substances one finds iridomyrmecin **98**, whose structure was determined by Fusco *et al.* (1955). This lactone is both an insecticide and an antibiotic. Cavill *et al.* (1956) also identified isoiridomyrmecin **99** in the Australian species *Iridomyrmex nitidus.* These authors also investigated the venom of *Myrmecia gulosa* which is essentially protein and from which electrophoresis yields histamine, hyaluronidase and a haemolytic factor (Pavan, 1975).

Iridodial **100** (Pavan & Trave, 1958) is a dialdehyde found in many species: *Tapinoma nigerrimum, Iridomyrmex rufoniger, I. nitidiceps, Dolichoderus scabridus,* etc. (Weatherston, 1967). Dolichodial **101** (or anisomorphal), which has a lachrymatory effect in vertebrates, is a dialdehyde isolated from *I. rufoniger, I. mymecodiae, D. scabridus, D. dentata, D. clarki* among others (Cavill & Hinterberger, 1961). At the same time, this substance was isolated from the phasmid insect *Anisomorpha buprestoides* (Meinwald *et al.,* 1962). Cavill and Hinterberger (1960) published the results of a systematic study of these terpenes in several ant species and have shown the existence of aliphatic ketones: methylheptenone, methylhexanone, propyl-isobutylketone. These terpenes can be found as mixtures of various stereoisomers and some have been synthesized (see the review by Weatherston, 1967). Pavan isolated dendrolasin **102** from the mandibular glands of the ant *Dendrolasius fuliginosus*; this substance is also believed to have insecticide properties (Quilico *et al.*, 1957) – 10 kg of this ant gave 70 ml of a liquid containing 75 per cent dendrolasin. This chemical has been found in several plants (Sakai *et al.,* 1965). According to Cavill *et al.* (1963, 1965), citral could be the biosynthetic precursor of these terpenes in insects (after reduction to citronellal, oxidation and cyclization to iridodial; the latter is thought to be the stepping stone for synthesis of iridolactones and dolichodial). Citral has been found in the mandibular glands of the ants *Atta sexdens rubropilosa* and *Acanthomyops claviger* (Butenandt, 1959 a, b; Weatherston, 1967). *De novo* biosynthesis of isoprenoids from acetate has been demonstrated for several insects (Meinwald *et al.,* 1966; Gordon *et al.,* 1963; Castellani & Pavan, 1966; Waldner *et al.,* 1969). Sakai *et al.* (1959) found some isoiridomyrmecin in a plant. These results seem to indicate the possibility that these substances can be synthesized by plants as well as by insects.

In the case of ants, the possibility that there are two sources of defence substances must be considered, namely the anal and the mandibular glands. The leaf-cutting and fungus-growing *Atta sexdens* synthesizes a fungicide, myrmicacin **103** which includes phenylacetic and indolacetic acids. These acids are constantly sprinkled in the nest. Myrmicacin prevents spore germination of fungus species other than those cultivated (20 mg of myrmicacin protect 100 g of plum flesh against fungal growth for at least three weeks). Phenylacetic acid is thought to prevent bacterial growth and indolacetic acid to favour mycelium development. The harvesting ant *Messor barbarus*, which collects pollen grains and seeds, also contains myrmicacin which, because of its anti-germinating powers, prevents the germination of grains and seeds stored in the nest (Schildknecht, 1971).

The list of benzoquinone-secreting insects is long and an outline of it can be found in Weatherston's review (1967). In many instances, quinones are mixed with hydrocarbons or with aliphatic aldehydes. These diluting agents might promote the penetration of these toxic chemicals through the cuticle. Ethylbenzoquinone **94** and toluquinone

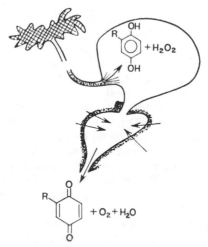

Fig. 9 'Reaction chamber' of the coleopteran *Brachynus*.

92 of the flour beetle *Tribolium* are responsible for the pink colour of flours: this is achieved by a chemical reaction with protein amine groups. According to Alexander and Barton (1943) these quinones might be of importance in limiting the growth of flour beetle colonies in flour (spermicidal action). Certain Carabidae also synthesize

Table 2a   Steroids in the prothoracic glands of water-beetles (Schildknecht, 1971)

Pregnene derivatives

R_1	R_2	R_3	*Steroid*	*Beetle* Quantities assayed in μg/animal
– C — CH$_2$OH $\parallel$ O	H	H	4-pregnene-21-ol-3, 20-dione	*Cybister limbatus* 1000 *fabricius* *C. lateralimarginalis* 3 *Dytiscus marginalis* 400 *Acilius sulcatus* 19 *Agabus bipustulatus*
– CHOH – CH$_3$	H	H	4–pregnene-20α-ol-3-one	*A. sulcatus* 1 *D. marginalis*
– CHOH – CH$_3$	H	H	4-pregnene-20β-ol-3-one	*Ilibius fenestratus* 1
– CHOH – CH$_2$OH	H	H	4-pregnene-20β, 21-diol-3-one	*I. fenestratus* 4
– CHOH – CH$_3$	H	OH	4-pregnene-15α, 20β-diol-3-one	*Platambus* 7 *maculatus* *I. fenestratus* 7
– C – CH$_3$ $\parallel$ O	OH	H	4-pregnene-12β ol-3, 20-dione	*C. lateralimarginalis* 28
– C – CH$_3$ $\parallel$ O	–O–C–CH$_2$–CH $=$ CH $\parallel$ O CH$_3$	H	4-pregnene-12β-ol-3-ol, 20-dione-pentenate-3	*C. lateralimarginalis* 4

benzoquinone **91** and toluquinone **92** for defence purposes. In other insects of the same family, the molecules used are sometimes of a very different nature: *Broscus cephalotes* uses salicylic aldehyde, *Platynus dorsalis* uses methyl salicylate, metacrylic and tiglic acids are found in *Calosoma sycophanta,* and also aliphatic ketones, phenols, etc. (Schildknecht, 1971). The Bombardier beetle *Brachynus crepitans* has a unique defence mechanism: Schildknecht (1971) notes that Pastor Wilhelm, who described the phenomenon in 1796, notices a sound similar to that of gun-shot, the smell of gunpowder and some heat. It is now known that the quinones in this beetle are only formed when needed. They are stored in the form of hydroquinones in a collecting bladder which also contains hydrogen peroxide (Fig. 9). A muscle allows these chemicals to enter an explosion chamber whose walls secrete enzymes (catalases and peroxidases) which break down hydrogen peroxide and oxidize hydroquinones to quinones (Schildknecht & Holoubeck, 1961). The temperature at the time of the explosion, measured with thermocouples or thermistors, is close to 100 °C. (Aneshansley *et al.,* 1969).

Since they live in water, Coleoptera of the genus *Dytiscus* require protection against micro-organisms which, by growing on their integument would tend to make them hydrophilic. These insects have glands (pygidial glands) on each side of the anus and, with the aid of their hind legs, the animals spread the secretions of these glands onto

Table 2b Steroids in the prothoracic glands of water-beetles (Schildknecht, 1971).

Pregnadiene derivatives

R_1	R_2	R_3	*Steroid*	*Beetle* Quantities assayed in μg/animal
– CO – CH₂OH	H	H	4, 6-pregnadiene-21-ol-3, 20-dione	*Cybister tripunctatus* 1
				C. lateralimarginalis 90
				Acilius sulcatus 56
– CHOH – CH₃	H	H	4, 6-pregnadiene-20α-ol-3-one	*C. lateralimarginalis* 140
				A. sulcatus 7
				Dytiscus marginalis
– CO – CH₃	H	H	4,6-pregnadiene-3, 20-dione	*A. sulcatus* 6
– CO – CH₃	H	OH	4, 6-pregnadiene-12β-ol-3, 20-dione	*C. lateralimarginalis* 6
				C. tripunctatus 1 050
– CHOH-CH₃	H	OH	4, 6-pregnadiene-12β. 20α-diol-3-one	*C. lateralimarginalis* 36
– CO – CH₃	OH	H	4, 6-pregnadiene-15α-ol-3, 20-dione	*Agabus sturmi*
– CO – CH₃	– O – CO – CH (CH₃)₂	H	4, 6-pregnadiene-15α-ol-3, 20-dione-isobutyrate	*Agabus sturmi*
– CH – CH₃ \ O – CO – CH (CH₂)₃	OH	H	4, 6-pregnadiene-15α-20β-diol-3-one-20-isobutyrate	*Agabus sturmi*

their body and the extremities of their wings (Schildknecht, 1971). These substances consist of aromatic acids and phenols among which are found benzoic acid, *p*-hydroxybenzoic acid, *p*-hydroxy-methyl benzoate, hydroxyquinone, etc. In the prothoracic glands *Dysticus* stores poisons which are very active in vertebrates, and in particular in fish, which are its prime enemies. These produce steroids deriving from pregnane (see Table 2). *Cybister limbatus fabricius* contains up to 1 mg cortexone. These steroids are found both in Dysticinae and in Columbetinae. In *Ilibius,* in addition to the steroids described in Table 2, testosterone, dehydrotestosterone, oestradiol and oestrone have also been found (Schildknecht, 1970, 1971).

The Gyrinidae, other aquatic coleopterans, which can be seen swimming around on the surface of still water, will release, when touched, a smelly secretion originating from the pygidial glands. This secretion is active against fish. Meinwald *et al.* (1972) identified the active product: (E,E,E) 3,7-dimethyl,8,11-dioxo-2,6,9-dodecatrienal (*Gyrinis ventralis, Dineutes hornii, D. serrulatus*). This structure was simultaneously described by Schildknecht *et al.* (1972) using *Gyrinus natator.* From these secretions, the authors obtained 3-methyl butanal. Gyrinidal, nor-sesquiterpenic aldehyde, is the most oxidized acyclic sesquiterpene known at the moment.

Paederus fuscipes is a European member of the Staphylinidae from which Cardani *et al.* (1965) isolated pederin **105**. This substance is a powerful blistering agent which makes it difficult to handle. It is known that cantharidin **106** from the Meloidae also causes blisters. It is sometimes used as an aphrodisiac but can be lethal in a dose of 10 mg (Weatherston, 1967).

The grasshopper *Romalea microptera* ejects, through the trachea, a predator-repulsing foam which contains an allenic sesquiterpene which probably derives from a plant carotenoid, phenols, *p*-benzoquinone and 2,5-dichlorophenol (Meinwald *et al.*, 1968 a; Eisner *et al.*, 1971). Chlorophenol is very active when assayed on ants. The authors suggest that this chlorinated derivative has its origin in insecticides absorbed by the grasshoppers.

The cantharid *Chauliognathus lecontei*, a soldier-beetle living on

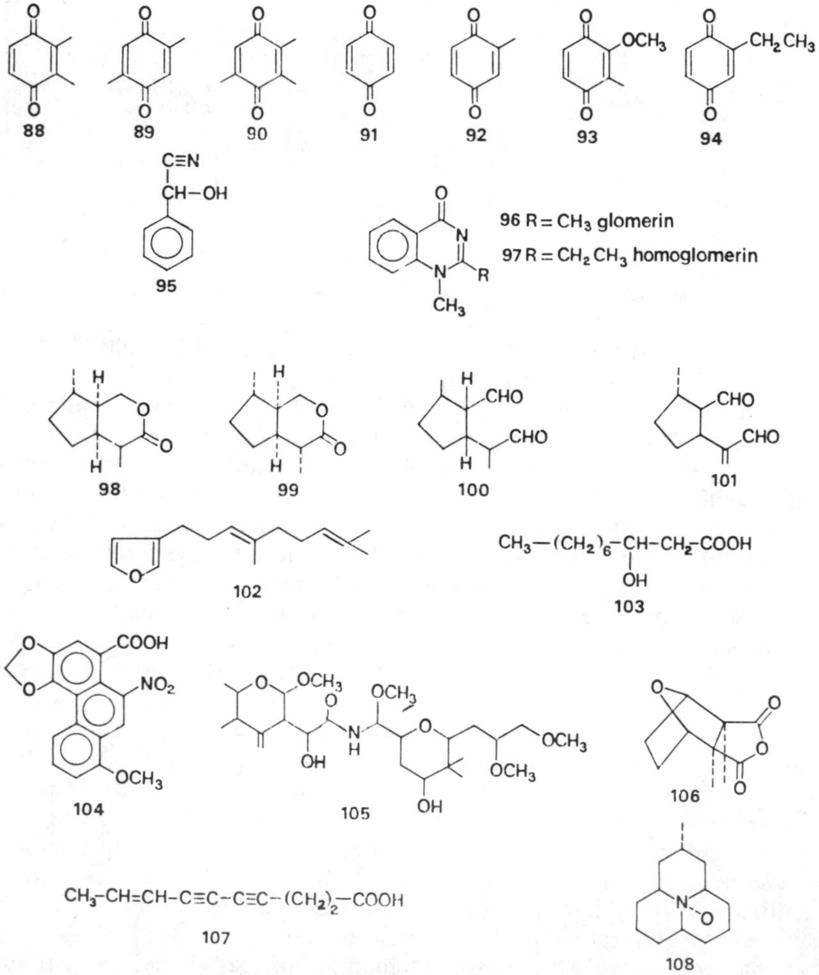

Fig. 10   Defence substances in insects.

Compositae seems to amass, for its defence, acetylenic acid (matricaria) **107** (Meinwald *et al.*, 1968 b).

The soldiers of various termite species secrete substances which rapidly solidify in air and which, because they can be thrown at a distance, cover enemies which have been reached. Benzoquinone and terpenes, which might have a toxic effect, have been found in these substances (Moore, 1968). The composition of the resin is not well known and may consist of mucopolysaccharides. In *Nasutitermes*, soldiers have a cephalic growth enabling them to throw the resin onto the enemy. This resin immediately solidifies forming strings (Eisner, 1972).

Vrkoc and Ubik (1974) determined the structure of the defensive substance found in the frontal gland of the Cuban termite soldiers *Prorhinotermes simplex*. It consists of a nitrated derivative which is quite rare in nature: 1-nitro *trans*-1 pentadecene.

The work of Tursch *et al.* (1971) and Karlsson and Lorsman (1972) has demonstrated that ladybirds are protected from birds by very bitter alkaloids (coccinellin **108**, adelin, etc.) These alkaloids are released together with the haemolymph during reflex haemorrhage.

The larva of the sow-fly, *Neodiprion sertifer* feeds on pine trees, *Pinus sylvestris* in particular. When it is distressed, it ejects by the mouth a drop of resin which will be used to smear the opponent. This resin is identical to that produced by the host plant. It is stored in two compressible pockets. This secondary use of a chemical weapon of the host by a carnivorous insect is quite remarkable (Eisner *et al.*, 1974).

Vertebrate toxins

The word toxin (*toxicon*) was originally used to describe extracts which were used to prepare poisonous arrows. The meaning of this word became obviously restricted, especially following its use to characterize certain bacterial proteins. It is often prefixed as in the words myco- and phytotoxins which have been mentioned earlier.

When toxic substances produced by vertebrates are considered the word venom is obvious. Its use in the description of the very important group of snake venoms became generalized. The word toxin is still found, as in batrachotoxin. It consists of a steroid alkaloid **109** isolated from South-American frogs with which Indians prepare arrow poisons (Tokuyama *et al.*, 1972; for a discussion of the definitions of toxins, see Cheymol *et al.*, 1970; Vogt, 1970). The LD_{50} for batrachotoxin in mice is 2 μg/kg. It is produced by *Phyllobates aurotaenia*. Other frogs, such as those of the families Dendrobatidae and Atelopidae possess pumiliotoxins (which include pumiliotoxin C **110**; Daly *et al.*, 1969).

The skin of another form of the *Dendrobates* genus, *D. histrionicus*, which lives in Colombia and is also used for arrow making, secretes an alkaloid series. These substances affect sodium and potassium transport across membranes and thus nerve impulse transmission. Six alkaloids have been identified, including histrionicotoxin, and spiropiperidin which has two branches with acetylenic groups. The other compounds are derived from the latter by their degree of saturation, and by the transformation of an acetylenic group into allene, etc. (Tokuyama *et al.,* 1974).

Many venomous vertebrates carry specialized glands containing highly active venoms as well as injection mechanisms, and lead an extrovert life. On the other hand, toxic animals which do not possess an injection mechanism usually lead a passive and hidden life. In recent years, published articles concerning venoms have numbered more than 10,000, which gives some indication of the current interest in the subject (Zlotkin, 1973).

The first mention of venoms is found in a text of an Egyptian papyrus of approximately 1600 B.C. (Leake, 1968). More recently, an excellent review laid the foundations of much present-day research, namely that of Phisalix (1922). A three volume review published recently is that of Bücherl *et al.* (1968).

In the following paragraphs we shall limit our review to examples of particular interest.

Since we started with amphibians we shall describe toad poisons, which are mixtures of biogenic amines such as adrenaline, noradrenaline, indolalkylamines (bufotenins) and bufotoxins. These products are synthesized by parathyroid glands and by the skin. The dried skins of toads have been used in Chinese medicine. Bufotoxins are combinations of bufogenins (such as bufotalin **111**) and suberylarginine. Bufotalin is part of a group of cardiotoxic steroids, many of which have been found in plants; some indeed are found in both plants and toads, such as hellebrigenin which is found in hellebore. Toad venoms, as well as those of some frogs, also contain proteins which act on blood pressure and which are haemolytic, etc. (Deulofeu & Ruveda, 1971; Meyer & Linde, 1971; Anastasi *et al.,* 1964, 1968, 1971). Salamanders secrete through the skin a series of biologically active substances such as biogenic amines, proteins, steroids and alkaloids. *Salamandra maculosa* contains the steroidic alkaloid samandarin **112**. Research concerning the tarichotoxin of North-American salamanders (*Taricha torosa, T. rivularis*) has established that it is identical to tetrodotoxin **113** (Habermehl, 1969, 1971; Deulofeu & Ruveda, 1971; Daly & Witkop, 1971; Meyer & Linde, 1971).

Tetrodotoxin **113** has been obtained from the ovaries and liver of the fish *Sphoeroides rubripes* (family Tetraodontidae). In spite of a very strict control of the culinary preparation of 'fugu', this fish still causes deaths in Japan. Tetrodotoxin blocks the movement of sodium across nerve and muscle membranes, thus inhibiting transmission of the impulse following excitation. Batrachotoxin causes membrane depolarization by increasing movement of sodium and is therefore a tetrodotoxin antagonist. The mode of action of the latter is therefore similar to that of saxitoxin **87**.

The chemical structure of tetrodotoxin is rather unique: it consists of an amino-perhydroquinazoline with a guanidine residue. This structure has been independently established by four different research groups: Goto *et al.*, 1965; Mosher *et al.*, 1964; Tsuda, 1966; and Woodward, 1964.

Fish toxins which originate in specialized glands usually consist of protein material. Fish displaying intermittent toxicity ought to be considered separately. Variation can depend on the seasons of the year or on geographical parameters and consumption of these fish is responsible for many cases of intoxication, particularly in the tropics (Halstead, 1965). It seems that these toxins are synthesized by certain types of algae and that they tend to accumulate in fish feeding on them. The illness known as 'ciguaterra' can therefore be caused by more than one fish species, but the proximity of coral-reefs seems to be a common factor. Nevertheless it appears that the chemical ciguatoxin is quite widespread (see the review by Banner, 1974).

Pahutoxin **114**, which is found in the mucus of *Lentiginonus*, is a choline ester of 3-acetoxy-hexadecanoic acid (Boylan & Scheuer, 1967).

Reptile venoms consist of polypeptide, protein and enzyme mixtures (Boquet, 1966; Lee, 1972; Zlotkin, 1973). There is only one lizard family among which venomous members can be found: the Helodermatidae. Their venom contains hyaluronidase and phos-

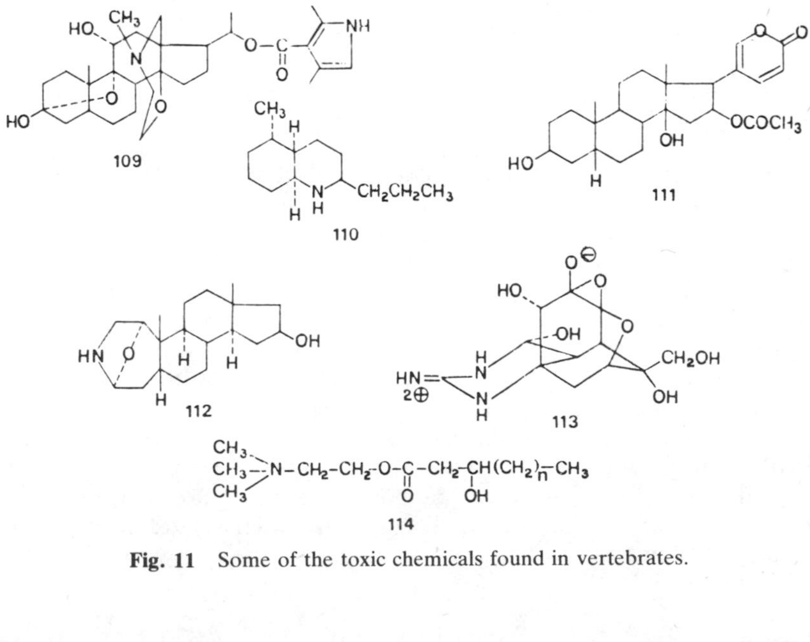

Fig. 11 Some of the toxic chemicals found in vertebrates.

pholipase A but no specific protein toxin. Venoms of the snake families Hydrophiidae (sea-snakes), Elapidae (cobras, mambas, etc.), Viperidae and Crotalidae contain many proteolytic enzymes: hyaluronidase, phosphoesterases, nucleotidase, cholinesterase, phospholipase A, esterases, exo- and endopeptidases, enzymes causing blood coagulation, etc. Venoms of the Viperidae and Crotalidae are particularly rich in proteolytic enzymes; that of Elapidae on the other hand consists to a large extent of esterases. Snake venoms represent some of the most concentrated sources of enzymes to be found in nature (Mebs, 1973). The Hydrophiidae have little or no enzyme in their venom, which contains neurotoxins consisting of basic polypeptides with 60 to 74 amino-acid residues with curarizing effects. Similar neurotoxins are to be found in the Elapidae. Yang *et al.* (1969) established the first snake toxin structure – cobrotoxin obtained from *Naja naja atra*. Its four disulphide bridges between cysteine residues are the cause of the biological activity. It is thought that alkaline amino-acids are responsible for selective binding to the receptor. Tryptophan and tyrosine residues contribute strongly to the conformation of cobrotoxin molecules that show the maximum efficiency.

The male duck-billed platypus (*Ornithorhyncus anatinus*) can inject venom using a spur found on the side of its posterior limbs. Painful but non-lethal accidents have been reported in man. The venom is however, lethal to small animals. Its constitution is still unclear but it is thought to act as a coagulating agent (Calaby, 1968).

Insectivorous shrews of the genera *Solenodon, Neomys* or *Blarina* have neurotoxins in submaxillary glands which they inject by biting. These toxins can kill small prey but in man they only cause a persistent and painful local reaction (Pournell, 1968; Pucek, 1968).

Mimicry

Mimicry is an important phenomenon indirectly related to chemical interactions between living organisms. To possess a chemical weapon is certainly an advantage from the evolutionary point of view. If the organism must first be swallowed by its predator in order to intoxicate it, this evolutionary progress benefits the species at the expense of the individual; protection occurs by backlash and can only be regarded as partially effective. The projection or the injection of venoms requires an extroverted way of life which exposes the animal to chance encounters and to the opponents' responses. A snake which has just bitten is often killed by the victim's companions. Another sort of evolutionary advantage consists of signalling that a weapon is available in order not to risk one's life by having to make use of it and still obliging others to keep at a distance. In order to relate mimicry interactions in terms of chemical defence, a definition will be briefly attempted.

It is a simplistic belief that animal behaviour consists essentially of fight, flight, coexistence or dissimulation. It is obvious that a green phytophage insect is protected from its predators because it blends with its most frequent backdrop. The green colour in Orthoptera is caused by accumulation of the chromoproteins of IX α-biliverdin **115** (Vuillaume, 1968; Rudiger *et al.*, 1969). The light-sensitivity of this pigment allows changes in colour caused by progressive oxidation. Discoloration can be complete or can result in the replacement of these pigments by more stable ones such as ommochromes. In the *Mantis* genus, the green colour can be obtained again during moulting if conditions have been favourable (Vuillaume, 1968). In *Oedipoda cerulescens*, visual feedback is responsible for adaptation to the environmental colours (Levita, 1966). This hiding behaviour or mimicry homochromy can be found in conjunction with homotypy: in these cases, the animal shape mimics that of an element in its environment. In the Lepidoptera, particularly in the larval stages, homochromy is often brought about by a pigment more stable than IX α-biliverdin **115**, pterobilin or IX γ-biliverdin **116** (Rudiger *et al.*, 1968; Vuillaume *et al.*, 1970).

In certain lepidopteran species, pterobilin is destroyed during imaginal moulting and is not found in the imago. If it is conserved, it leads to blue or green butterflies. Blue pigments deriving from pterobilin (neopterobilins; Choussy & Barbier, 1973) have been isolated from some species. Pterobilin is obtained by a non-classical opening of IX protoporphyrin at the γ-methene bridge level.

Pigments used by insects belong to extremely varied families: melanins, ommochromes, pterins, carotenoids, flavones, biliary pigments, etc. They are sometimes excretions with or without adaptative roles. 'The mechanism by which a mimicry analogy is produced in nature is a problem which involves that of the origin of the species and all adaptations' (Bates, 1862, cited by Vuillaume, 1969). A wide variety of chromatic adaptations is found in animals. They are not always aimed at camouflage but are sometimes related to particular behaviour.

The egg-shells of the quail are coloured with IX α-biliverdin and IX protoporphyrin. A final coloration covers the egg surface with protoporphyrin spots, whose number, size and shape are thought to be genetically determined (Lucotte, 1974; Lucotte *et al.*, 1974).

Table 3 summarizes some of these adaptations taking into account the suggestion of Cott (1938) and of Huxley (1934). The usefulness of this classification is essentially the language it makes available. For instance the term *aposematic colour*, which describes a colour which warns of danger, is often used. Between 1849 and 1860, the British naturalist Henry W. Bates crossed the Amazon basin looking for new Lepidoptera species. In 1862 he described the occurrence of Pieridae which were imitating Heliconidae because the latter are protected from birds by bitter substances originating in plants on which the larvae feed (Wickler, 1968). Batesian mimicry was born.

Table 3 Chromatic adaptations (according to Caillois, 1963).

POULTON-COTT			Julian HUXLEY
Apatetic colours (*misleading*)	cryptic (*dissimulation*)	procryptic (*protection*) / anticryptic (*attack*)	cryptic colours (*dissimulation*)
	pseudosematic (*misleading warning*)	pseudoaposematic (*false danger, deceit*) pseudoepisematic (*bait*) parasematic (*to divert*)	phaneric colours (*true or misleading signal*)
Sematic colours (*true signal*)	aposematic synaposematic (*to repel*) episematic (*to attract*)		

IXα-Biliverdin 115

Pterobilin (IXγ-biliverdin) 116

Maximum protection of a species is reached by the conjunction of factors such as bad taste, a vivid coloration or combinations of colours (red, black, yellow, etc.) and, for instance, a low motility. Butterflies which are easy to catch are often abandoned by birds if they are bitter or if the bird associates bitterness with the colour. Later Müller attempted to demonstrate the uniformity of aposematic signals within a single insect family and also among related species in order to achieve maximum efficiency: e.g. generalization of yellow and black banding in members of the Hymenoptera which are protected by a sting: Müllerian mimicry. It is tempting to interpret these observations in a finalistic manner by concluding in terms of the 'unity of purpose of the species' (Caillois, cited by Vuillaume, 1969).

Aposematism can be linked to group phenomena which bring into play pheromones (Eisner, 1970) or auditory signals (Haskell, 1961). A discussion of the various forms of mimicry will be found in a review by Leroy (1974). This author classifies animals in terms of their response to light. This procedure displays a factor which is essential in the understanding of animal behaviour.

To Batesian mimicry, which links visual signals with a chemical effect (bad taste, toxicity, venom, etc.) and Müllerian mimicry, must be added Mertesian mimicry (the copying of a species displaying Batesian mimicry by a harmless, related species), and Peekhamian mimicry (or aggressive mimicry, in which the predator imitates the harmless model or a bait). The bitter substances associated with asposematic signals usually originate in the plants on which the insects feed, such as cardiotoxic steroid glycosides, but exceptions are found such as the alkaloids isolated by Tursh *et al.* in ladybirds (see above).

Mimicry is one of a number of phenomena of a very complex nature for which attempts at understanding are often limited by our fragmentary knowledge. Examples of behaviour which does not fall easily into predetermined classes are numerous. Interpretation is confused by anthropomorphic projections which sometimes appear insidiously in deductions which at first hand seem quite reasonable.

The conclusion drawn by Leroy (1974) according to which 'mimicry arises out of the domain of expression' underlines an essential need of living matter; to act by transmission of messages. This need makes use of nearly all means: sound, light, shape, molecules, etc. either singly or in combinations. The existence of such combinations renders the science of chemical ecology that much more complex.

Bioluminescence

Bioluminescence, or the emission of visible light by living organisms, is a phenomenon particularly widespread in marine life. It can be caused, such as in the case of fish, by symbiotic bacteria, but other animals possess an original photon-producing mechanism. Such animals are found among the Radiolaria, Peridinia, Hydrozoa, Ctenophora, Nemertea, Polychaeta, Oligochaeta, Mollusca and the Crustacea. With the exception of *Latia*, a small mollusc found in New Guinea, bioluminescence is not found in fresh water. On land, some fungi and some insects nevertheless have this ability. In some cases, interspecific interactions are quite obvious, as in the case of the fish *Lophius piscatorius* which has a lantern at the end of a tentacle which acts as a bait, or alternatively intraspecific interactions (e.g. the firefly). However, in many cases the role of luminescence is difficult to assess. In the fish *Anomalops,* a hinged muscle which is found at its anterior end can rotate by 180° the luminescent organ in which bacteria are found, thus achieving directional light emission under voluntary control. *Odontosyllis,* the 'fireworm' of Bermuda, releases luminescent material in the water at reproduction time. The crustacean *Cyprinida hilgendorfii* found in the sea of Japan releases for defence purposes a cloud of luminescent material and thus deceives its enemies. This secretion originates in two glands, one of which contains a luciferin and the other a luciferase enzyme. The overall mechanism for the reaction is as follows:

Luciferin + Luciferase + O_2 → (Intermediate in excited state) →
        Intermediate + LIGHT

The work of Kishi *et al.* (1966) led to the determination of the structure **119** for luciferin in this crustacean. The accepted mechanism relies on the formation of a hydroperoxide intermediate (due to the mobility of the α-carbonyl proton) (McCapra, 1973; Shimomura & Johnson, 1971).

Visual acuity of the inhabitants of marine depths is limited to that of abstract pictures which nevertheless are absolutely recognizable. Carnivorous animals often resort to deception as a means of attracting prey. About 150 species of fish with lanterns have been observed and can be classified according to the luminous pictures they project. Members of the Myctophilae family produce dazzling flashes which momentarily daze the aggressor. The cuttlefish *Sepiola ligulata, Rondeletiola minor* and *Heterotheutis,* when in danger, can empty their stock of fluorescent bacteria colonies to form luminous clouds while they themselves become invisible.

Firefly signals are sometimes very complex: their purpose is species recognition. In *Photimus pyralis,* a North-American species, the male produces flashes at irregular intervals to which the female of the species answers. The liberation of chemicals responsible for these flashes is under nervous control. In fireflies, ATP and magnesium ions are required for the luciferin–luciferase reaction and in the species mentioned above the efficiency in terms of light energy emission is 88 per cent. The first stage of the mechanism is the formation of acyl-adenylate **121** from luciferin **120**. Here, luciferase has a double purpose: firstly it is an adenyl carrier and secondly it acts as an oxidizing agent in the presence of atmospheric oxygen. D-cysteine is required in luciferin **120** biosynthesis; a synthetic compound obtained with L-cysteine is not used by the enzyme although *in vitro* it is still capable of liberating light (White *et al.,* 1963). Our knowledge of the precise mechanism by which photons are liberated has improved considerably since the use of synthetic models (McCapra *et al.,* 1968; Hopkins *et al.,* 1967; Suzuki *et al.,* 1969). The very high sensitivity of the reaction to ATP, oxygen and calcium ions led to very precise assay methods for these molecules (Ashley, 1971; McElroy & Strehler, 1954). Females of various *Photuris* species catch (and eat) the male representatives of related species by tuning their luminous response to their signals. This phenomenon suggests a very complex nervous control over luminescence (Lloyd, 1975).

The comment of McCapra (1973) concerning the future development of bioluminescence studies is of interest: 'The extraordinary use of light made by bioluminescent organisms opens up an exciting area for ecologists and theoreticians of evolution.' However, in spite of the fact that several thousand species have the ability to emit light, interpretations in terms of evolution remain difficult and hazardous. For example, quite a large number of luminescent species are found among the hydroids and jelly-fish; on the other hand sea-anemones and corals do not show this property. In some cases, the use of lanterns is of a quite obvious advantage, for example in the attraction of prey towards nematocysts, but in other cases the evolutionary advantage of light emission is quite obscure.

In the jelly-fish *Aequorea aequorea,* the light-emission mechanism is very different from those described above. It is a photoprotein that is active and responsible for the blue light of the photophores in the coat. Oxygen is not required, only calcium ions (Ashley, 1971). This mechanism is found in other coelenterates: *Obilia, Pelagio, Mnemiopsis, Renilla* and *Lovenella* (Shimomura & Johnson, 1972). The luciferin in *Aequorea* is thought to be identical to that found in *Renilla* (Cormier *et al.,* 1973, 1974). The structure of the chromophore is difficult to determine because of the small quantities which can be obtained from these animals. Forty thousand *Renilla* specimens are needed to provide only 0.5 mg pure luciferin. An attempt at synthesizing these chemicals has been made and compound **117** led to a luminous yield comparable to that of natural luciferin. In *Renilla* the latter is found as an inactive sulphate and the reactions leading to photon emission are as follows:

Luciferyl-sulphate $\rightarrow$ PAP (3'5'-diphospho adenosine) $\rightarrow$
Luciferin + PAPS (sulphokinase) luciferin + O_2 $\rightarrow$
Oxyluciferin + CO_2 + LIGHT

Luciferase is a protein with a molecular weight of approximately 24,000. The structure of oxyluciferin **118** has been determined through synthesis. A proposed mechanism is shown in Fig. 12 (review by Cormier *et al.,* 1974).

It would appear that all luciferins mentioned above derive from three amino-acids. The emission wavelength varies depending on the substitutions: tyrosine, for example, leads to luciferins with an ionizable group and to a wider emission waveband.

Latia neriloides, a small mollusc of New Guinea and the only fresh-water animal which is luminescent, possesses a fluorescent protein (purple protein) as a co-factor (Shimomura & Johnson, 1968). These authors established structure **122** for the corresponding luciferin. It differs from the luciferins already mentioned by the absence of nitrogen and by the remarkable presence of enol formate. The bioluminescence of fungi (*Armilla mellea* which grows on decaying wood) still remains a mystery. In bacteria, the classical reaction involving luciferin and luciferase does not exist. The co-factors are aliphatic aldehydes (maximum activity being found for

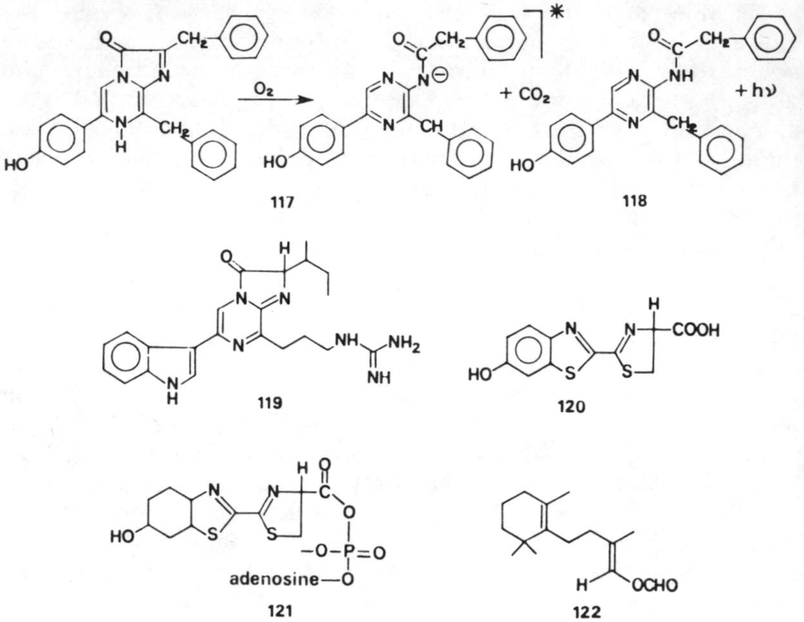

Fig. 12 Molecules taking part in bioluminescence processes.

chain lengths of 10 to 18 carbon atoms) (Shimomura *et al.*, 1972). $FMNH_2$ (dihydroflavine mononucleotide) is required (Hastings *et al.*, 1965) and the proposed general mechanism is as follows:

Luciferase + $FMNH_2$ ⇌ (Luciferase-$FMNH_2$)
Luciferase-$FMNH_2$) + RCHO ⇌ (Active complex)
$FMNH_2$ + O_2 → FMN + H_2O_2
(Active complex) + H_2O_2 → Luciferase + FMN + RCOOH + LIGHT

Vigny and Michelson (1974) demonstrated the oxidation of 1-^{14}C lauryl aldehyde into lauric acid by the luciferase of *Photobacterium sepia*. The bacterial luciferase can therefore be considered an aldehyde oxidase. The energy released by this oxidation is used by the emitter. These authors doubt that FMN is the flavine actually active *in vivo*, since a particular flavine has been discovered in luciferase.

Chapter 3

The ecological importance of sterols in invertebrates

The need for sterols illustrates a unifying aspect of the living world. One of the most important is cholesterol, which is both a cell constituent and a stepping stone for various metabolic pathways. Cells able to develop in the absence of sterols belong only to primitive systems such as bacteria. Thus, the discovery of cholesterol in plants and of steroid hormones in invertebrates was of great importance. Many invertebrates, such as insects, are not able to synthesize cholesterol which they nevertheless require. These animals must therefore look for food sources which contain it, or develop indirect means of obtaining it. In nature, cholesterol plays a fundamental ecological role. It is true, however, that there are other examples of molecules for which a constant search must be made by living organisms. Sterols with 28 and 29 carbon atoms which are common in plants are formed by methylation of the side-chains of their precursors derived from cycloartenol. In contrast to lanosterol, the classical intermediate between mevalonic acid and C_{27} sterols in vertebrates, cycloartenol in plants is the cyclization product of squalene and is the starting point for a number of steroid metabolites.

Hormonal control in insects

The existence of hormones in insects was first demonstrated by Sir Vincent Wigglesworth. The present-day knowledge of the field can be summarized as in Fig. 13. Two antagonistic hormonal systems control insect development: juvenile hormone is responsible for maintaining the larval state by favouring the differentiation of larval structures as opposed to imaginal structures. Furthermore, it is often indispensable for normal development in the imago itself. The moulting hormone (ecdysone) triggers moulting. Towards the end of larval life, this hormone acts by itself to allow the complete development of imaginal structures. It is known that it specifically induces protein biosynthesis by activating messenger RNA synthesis (reviews by Berkoff, 1971; Menn & Beroza, 1971). Cerebral hormone (ecdysiotropin) stimulates the release of moulting hormone from the prothoracic gland.

Juvenile hormone is secreted by the corpora allata and was isolated for the first time from the butterfly *Platysamia cecropia* by Williams

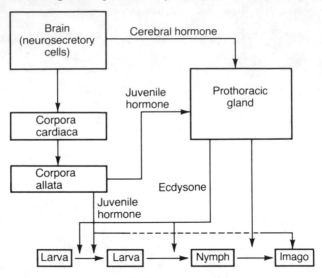

Fig. 13   Scheme for hormonal control in insects.

and Law (1965). Williams noticed as early as 1956 the insecticidal properties of this hormone. Structure **123** was obtained by Dahm *et al.* (1967, 1968) who also described its synthesis. Meyer *et al.* (1968) and Johnson *et al.* (1969) isolated a second juvenile hormone from *Cecropia* (**124**) which differs from the preceding one by the presence of a single ethyl branch. From another lepidopteran, *Manduca sexta,* Judy *et al.* (1973) obtained the juvenile hormones **124** and **125** (which do not have ethyl groups and are therefore methyl epoxyfarnesoates). Biosynthesis of juvenile hormones is thought to start from acetate and propionate, at least as far as the backbone of the molecule is concerned. Methyl ester is introduced by S-adenosylmethionine (Schooley *et al.*, 1973). Jennings *et al.* (1975) established that the ethyl group in the juvenile hormone of *Manduca sexta* comes from homomevalonic acid. This was the first example of a mevalonic acid homologue taking part in terpene biosynthesis. Farnesol and farnesal, which are of vegetable origin, have a similar, although considerably lower, activity (Schmialek, 1961). The condensation of hydrogen chloride with farnesoic acid yields a mixture of chlorinated isomers, which at a concentration as low as 1 in 100,000 are able to stop the development of mosquito larvae completely (Law *et al.* 1966). The dichloroester **126** (synthesized by Romanuk *et al.,* 1967) can, at concentrations as low as 1 μg per animal, induce irreversible sterility in the female of the species *Pyrrhocoris apterus*. When treated males mate with untreated females, they transmit by contact enough of the substance to completely sterilize the females' eggs (Manser *et al.,* 1968). Juvabione **127**, a constituent of the conifer *Abies balsamea,* which is used for paper pulp, displays remarkable activity ('paper

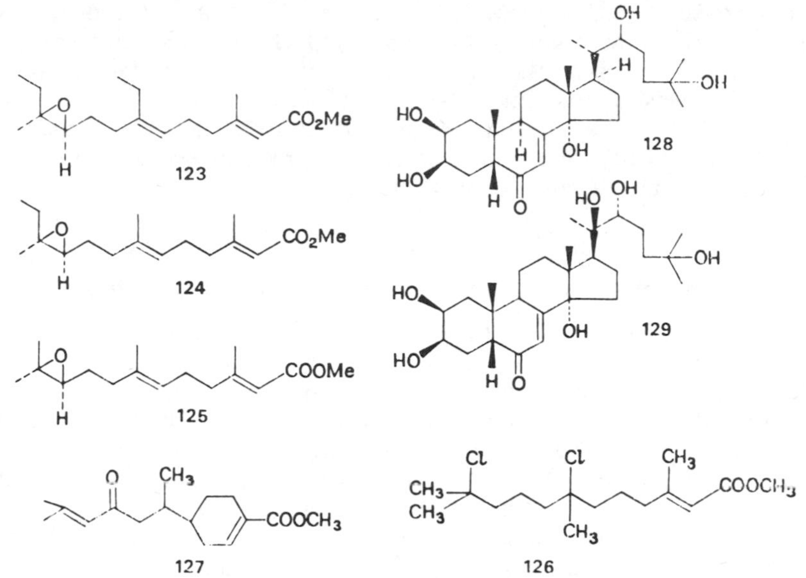

Fig. 14 Hormones of insects and mimicry.

factor') (Williams & Slama, 1966). According to these authors, extracts of the *New York Times* are very active while extracts of the London *Times* are quite inactive (see also Bowers *et al.*, 1966; Birch *et al.*, 1970)!

The determination of the structure of ecdysone **128** was achieved by Butenandt and Karlson in 1954; (25 mg of the substance was obtained from 500 kg of *Bombyx mori* pupae!) (Huber & Hoppe, 1965).

β-Ecdysone (also called 20-hydroxy β-ecdysone, crustecdysone or ecdysterone) has been found in *Bombyx mori* as well as in marine arthropods (Horn *et al.*, 1966). Nakanishi *et al.* (1966) discovered the plant ecdysones (for instance, crustecdysone obtained from the Australian tree *Podocarpus elatus*). Since then, many similar phytosteroids have been isolated from plants (Nakanishi, 1971). Syntheses made by Sorm and co-workers showed the possibility of having anti-ecdysones that derive from ecdysone (6-keto derivatives saturated in positions 7 and 8) (Velgova *et al.*, 1969). In *Calliphora*, cholesterol is the biological precursor of ecdysone (Karlson and Hoffmeister, 1963).

The discovery that insect hormones and analogous compounds with similar activities are found in many plant species is somewhat surprising, and their significance as far as evolution is concerned remains obscure.

Williams and Slama (1966) boldly suggested that conifers had developed, by means of juvabione, a sophisticated defence mechanism against insects which acts by mimicking a hormone which these animals need, but which, because of its very high concentration, leads to dramatic consequences. Galbraith and Horn (1966) proposed a similar hypothesis in the case of the tree *Podocarpus elatus*, which contains considerable amounts of crustecdysone.

In arthropods, the need for cholesterol is absolute because it is the precursor of ecdysones. This fact leads naturally to a consideration of the ecological relevance of this chemical.

The origin of cholesterol in insects and marine invertebrates

Cholesterol is an absolute requirement for insect life for two reasons. Firstly, it is the biological precursor of ecdysone, and secondly it is an essential growth factor which they cannot synthesize. In 1935, Hobson observed that *Lucilia sericata* needed this sterol for normal growth and Hoogt in 1936 showed the same requirement for the fruit fly *Drosophila melanogaster*. Many authors have since reported the effect of various sterols on insect development, particularly sterols in the C_{27}, C_{28} and C_{29} groups. Morisaki *et al.* (1974) have published a list of these compounds distinguishing between those that can be made use of by *Bombyx mori* and those which cannot (for reviews, see Clayton, 1964; Gilbert, 1967). The dietary requirements for cholesterol vary according to species, from 0.01 per cent to 0.1 per cent. In the case of *Locusta migratoria* toxic effects are observed for dietary intake of cholesterol greater than 1.4 per cent (Dadd, 1960). Phytophagous insects find in their diet mainly sterols with 28 and 29 carbon atoms such as β-sitosterol, campesterol, stigmasterol, etc. Cholesterol is found in many plants, sometimes in traces and sometimes in larger amounts (Johnson *et al.*, 1963; Heftmann, 1963; Devys & Barbier, 1965). Modification of the cholesterol structure by hydroxylation and reduction reduces the biological activity of the molecule (see Allais, 1974, for a review). Clark and Bloch showed that *Dermestes vulpinus* breeds normally on a poor cholesterol diet as long as other sterols are found in the food. This sparing effect has since been reported by other authors (Clayton & Edwards, 1961; Clayton & Bloch, 1963).

Clark and Bloch (1959) and Clayton (1960) observed that *Blatella germanica* has the ability to transform ergosterol into 22-dehydrocholesterol thus eliminating a carbon atom from the side-chain. This discovery initiated much research on the dealkylation of side-chains in phytosterols. In leaves, the predominant sterol is β-sitosterol, which insects convert into cholesterol and therefore is of great ecological significance (Robbins *et al.*, 1962; Svoboda & Robbins, 1967; Ritter & Wientjens, 1967; Allais *et al.*, 1971; Allais & Barbier, 1971; Allais *et al.*, 1973). The conversion is carried out via fucosterol, probably 24(28)-epoxyfucosterol and desmosterol (see pathway in Fig. 15). In the case of *Locusta migratoria*, sterols with 28 carbon atoms such as campesterol are not intermediates in the

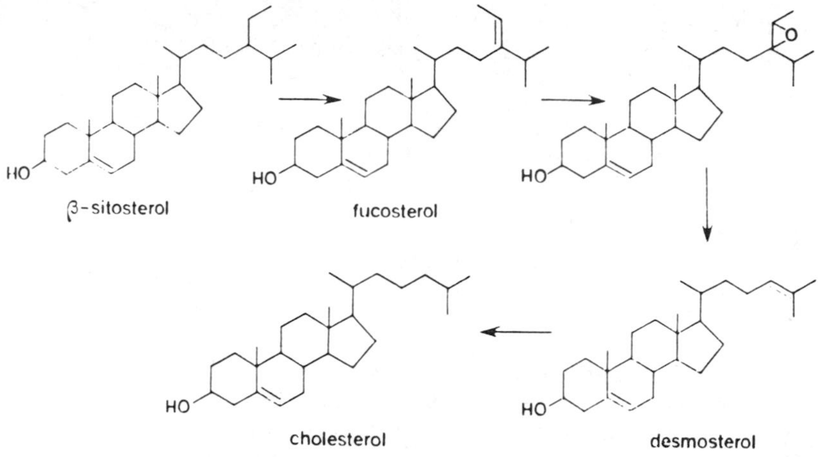

Fig. 15 Dealkylation of the β-sitosterol side-chain by insects.

pathway, in spite of the fact that 24-methylene cholesterol (Barbier, 1966) can be transformed into cholesterol.

As in the case of insects, crustaceans cannot synthesize cholesterol from acetate, and must find it either in their diet or obtain it by degradation of the side-chain of phytosterols. Molluscs on the other hand, including all gasteropods which have been investigated and some lamellibranchs (the mussel *Mytilus edulis* and the clam *Saxidomus giganteus*, for example) are able to synthesize cholesterol all the way from acetate. The first demonstration of the co-existence of the two different cholesterol origins (namely by complete synthesis and from phytosterol degradation) in a single animal, was in the limpet *Patella vulgata*. It is likely that this phenomenon will be observed in other gasteropods. In *Patella* the β-sitosterol degradation follows the same pathway as in insects, i.e. via fucosterol and desmosterol (Fig. 15) (Collignon-Thiennot *et al.*, 1973).

Cholesterol is certainly the 'best suited' (Bergmann) of all sterols and at the cellular level, cholesterol requirements are general. It is therefore intriguing that plant cells synthesize sterols with 28 and 29 carbon atoms, and somewhat surprising that some invertebrates such as insects are unable to achieve *in toto* cholesterol biosynthesis. We have formulated a hypothesis, according to which at a certain period of evolution, all animals had the ability to synthesize cholesterol from acetate and that this was progressively lost as a result of an energetically more favourable alternative, phytosterol dealkylation. The limpet case tends to support this hypothesis, for, in having both pathways, it is an example of an intermediary stage of this evolutionary process. A certain amount of caution is, however, required. When the number of existing animal species is considered, a very small percentage of them have been investigated in terms of their sterol requirements and in terms of phytosterol transformations (Zandee, 1966; Voogt, 1972; Collignon-Thiennot *et al.*, 1973).

The presence of steroids in invertebrates

As we have seen above, ecdysones are synthesized from cholesterol which is obtained either from a total synthesis starting from acetate (as in the gasteropods) or from phytosterol dealkylation (as in the arthropods). Aquatic Coleoptera secrete steroids which are the same as those commonly found in vertebrates. The biological significance of the presence of these substances is unclear. Progesterone, pregnenolone and dehydroepiandrosterone have been isolated from the flour beetle *Tribolium confusum* but progesterone is found as a trace substance in the flour which is the usual diet of this insect (Dubé & Lemonde, 1970). Progesterone is accumulated in the female part of the gonads of the scallop *Pecten* (Mollusca) and no explanation of this phenomenon can be found (Saliot & Barbier, 1971). In some invertebrates enzymes which are able to transform steroids (hydroxylases, reductases, oxidases, etc.) have been observed but they may be rather non-specific enzymes or even induced by the presence of the substrate. The role of intestinal flora is difficult to eliminate when the assays are carried out in the absence of antibiotics. Experiments carried out on the honey-bee *Apis mellifera* have led to some interesting observations: 10 to 16 per cent of progesterone or 17-hydroxy α-progesterone is transformed when these steroids are mixed in with the food and this is observed whether or not antibiotics are used. These steroids are not found in the bee in its natural environment. The metabolites which appear correspond to transformations brought about by hydroxylases, reductases and oxidases (Veith *et al.*, 1974). Echinoderms synthesize glycosides which are related to saponins: holothurins from holothurians and asterosaponins from starfish. Asterosaponin A has anti-hormonal activities (see Ch. 2 – Toxins in marine invertebrates).

It is likely that in the future more invertebrate steroids will be isolated but at the present time there is no evidence to suggest that these substances play a role similar to that in vertebrates.

It appears that in the course of evolution, all molecules are made use of in the best way possible whether they originate in the diet or are obtained by degradations caused by enzymes with little specificity.

Plant cholesterol

Cholesterol is usually found in large amounts in the diet of carnivorous invertebrates. In plants, however, it often exists in much lower quantities, but is still significant for plant eaters. The discovery of plant cholesterol is related to technical developments occurring since 1960. Before that time, zoosterols were known (in C_{27}) as well as phytosterols (in C_{28} and C_{29}) and this criterion was thought to biochemically differentiate plant and animal life. Indeed, the work of Tsuda that demonstrated the presence of cholesterol in red algae was believed to be the illustration of an exception to the rule. Now we know that cholesterol is widespread in plant life. It is even sometimes found in large quantities, for instance in the pollen of the Compositae or that of poplars (Devys & Barbier, 1965; Standifer *et al.*, 1968). It is however, always found in conjunction with the usual sterols having 28 and 29 carbon atoms, even in the case of the red algae (Alcaide *et al.*, 1968). Apart from its role, together with other sterols, in the constitution of cell membranes, it appears that in plants, cholesterol is

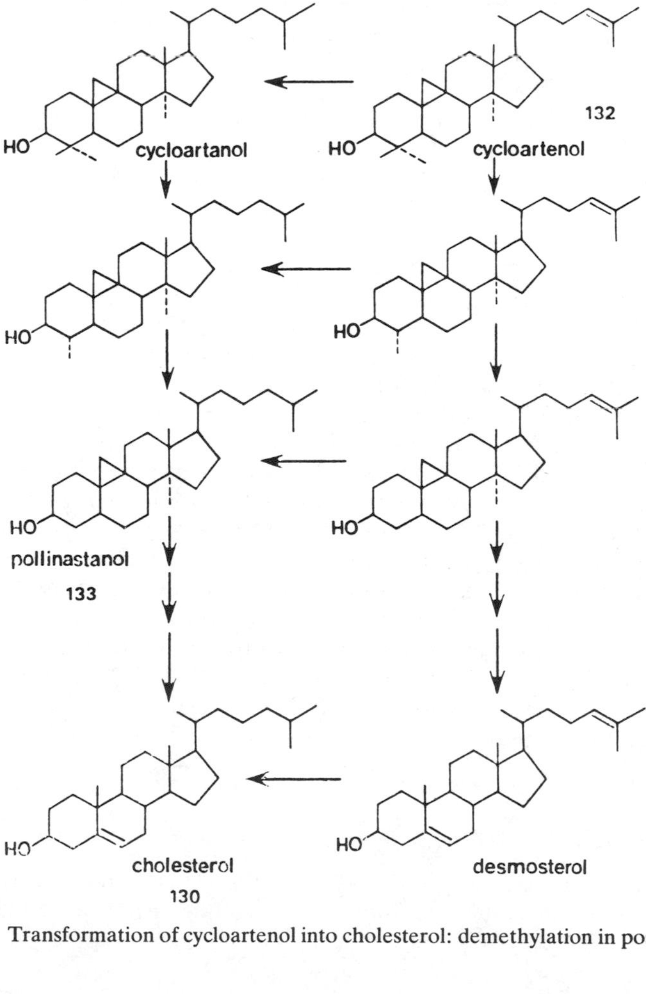

Fig. 16   Transformation of cycloartenol into cholesterol: demethylation in position 4.

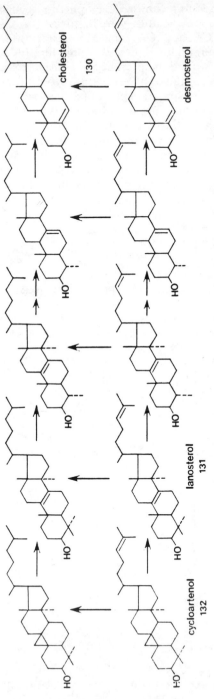

Fig. 17   Transformation of cycloartenol into cholesterol: early opening of cyclopropane ring.

the stepping stone for a whole family of metabolic pathways leading to the synthesis of a variety of sterols. It is the biological precursor of phytoecdysones (Hikino *et al.*, 1970; Heftmann *et al.*, 1968; de Souza *et al.*, 1970; Sauer *et al.*, 1968), of cardenolides (Tschesche & Lillienweiss, 1964), of steroid hormones, progesterone, pregnenolone, etc. (Bennett & Heftmann, 1965; Gawienowski & Gibbs, 1969). C_{21} steroids are precursors of a large number of steroid alkaloids (Tschesche, 1965) which might play a part in defence mechanisms against herbivorous animals because of their bitter taste, and the same function might be attributed to cardenolides.

Cholesterol **130** synthesis in higher plants occurs along the classical pathway from mevalonate to squalene. The triterpenic alcohol which is formed following cyclization is not lanosterol **131** but cycloartenol **132** (an exception is found in the latex of *Euphorbia*). In fungi, however, lanosterol is the rule and cycloartenol has never been found. Starting from these triterpenic alcohols (Lederer, 1969; Goodwin, 1970), cholesterol can be obtained, or even C_{28} and C_{29} sterols after biological methylations in position 24 by S-adenosylmethionine.

If cycloartenol **132** transformation starts by demethylations in position 4, cholesterol is obtained via pollinastanol **133**, according to the scheme shown in Fig. 16. In such cases, for instance in the fern *Polypodium vulgare* or in the sarsaparilla *Smilax medica*, pollinastanol is an important intermediate (Hügel *et al.*, 1964; Devys & Barbier, 1967; Devys *et al.*, 1969 a and b).

A second pathway starts by the opening of 19,9-cycloprane which leads to the usual intermediates found after lanosterol demethylation (Fig. 17).

C_{28} and C_{29} phytosterols

The study of plant sterols with 28 and 29 carbon atoms was carried out by various research groups simultaneously, in particular the groups of Goodwin, Ourisson and Lederer. The most widespread plant sterol is certainly β-sitosterol which is often found together with campesterol and stigmasterol (24α). In fungi, the most common sterol is ergosterol (24β). 24-Methylene cholesterol has been found in many plants and abundantly in pollens and unicellular algae. In brown algae the main sterol is fucosterol. In some plants, the main sterol is not Δ_5 by Δ_7 such as in the case of α-spinasterol in spinach (see the sterol lists in Tables 4, 5 and 6).

Methylation of the side-chains of cycloartenol has been observed: cyclolaudenol, cycloeucalenol (Clayton, 1965; Hewlins *et al.*, 1969; Goad, 1970; Devys, 1971; Alcaide, 1971). 24-Methyl and 24-methylene pollinastanol have been described (Doyle *et al.*, 1972; Knapp *et al.*, 1972; Rohmer & Brandt, 1973).

The introduction of further carbon atoms into side-chains is achieved by S-adenosyl-methionine on the unsaturated 24 of cycloartenol and of its derivatives. A family of compounds results with one or two extra carbon atoms in position 24 (depending on whether one or two biological methylations occurred), which have lost one or two methyl groups in position 4 (pollinastanol series) for which the cyclopropane has been opened up (return to the series of lanosterol and to lophenols), etc.

The various series which are successively obtained during these transformations are shown in Figs. 18 and 19. The schemes are really three-dimensional, since the two figures must be superimposed; thus the transformation from a substance in Fig. 18 to the corresponding steroid in Fig. 19 is possible. The number of possibilities is vast and the biological characteristic of each plant will determine the chosen pathways linking triterpenic alcohols, methyl sterols and sterols.

The function of cycloartenol as a substrate for biological

Table 4   Plant sterols.

methylations leading to C_{28} and C_{29} sterols had been postulated as early as 1965 by von Ardenne and his colleagues and demonstrated by Ourisson's group. Parallel work by Goodwin and his team led to similar results. Alcaide and Devys studied the incorporation of radioactive precursors and showed that biological methylations preferentially occur on triterpenic alcohols that are unsaturated in 24,25 or 24(28) rather than on the corresponding sterols such as desmosterol or 24-methylene cholesterol (Alcaide *et al.*, 1968, 1969).

The 5,6 unsaturation is introduced by means of the 5,7-diene and starts from Δ_7 steroid.

Double bonds in position 22 are formed after the methylations (brassicasterol, stigmasterol).

Table 5   Sterols and triterpenic alcohols in plants.

Table 6 Triterpenic alcohols and methyl sterols in plants.

List of sterols, methyl sterols and triterpenic alcohols shown in Tables 4, 5 and 6.

I	Cholesterol	XXII	Elasterol
II	22-dehydrocholesterol	XXIII	Cycloartenol
III	Desmosterol	XXIV	Cycloartanol
IV	7-dehydrocholesterol	XXV	31-nor-cycloartanol
V	β-sitosterol	XXVI	Pollinastanol
VI	Campesterol	XXVII	31-nor-cycloartenol
VII	Stigmasterol	XXVIII	24-methyl cycloartanol
VIII	Ergosterol	XXIX	Cyclolaudenol
IX	Brassicasterol	XXX	Cycloneolitsine
X	Poriferasterol	XXXI	31-nor-cyclolaudenol
XI	22-dihydrobrassicasterol	XXXII	Cycloeucalenol
XII	Clionasterol	XXXIII	24-methylene pollinastanol
XIII	24-methylene cholesterol	XXXIV	24-methyl pollinastanol
XIV	Fucosterol	XXXV	Lanosterol
XV	Sargasterol	XXXVI	31-nor-lanosterol
XVI	Isofucosterol	XXXVII	Obtusifoliol
XVII	α-spinasterol	XXXVIII	Lophenol
XVIII	Δ⁷ Stigmasterol	XXXIX	24-methyl lophenol
XIX	5-dihydro ergosterol	XL	Citrostadienol
XX	Chondrillasterol	XLI	Isomer in position 24 of XL
XXI	Cucurbitasterol		

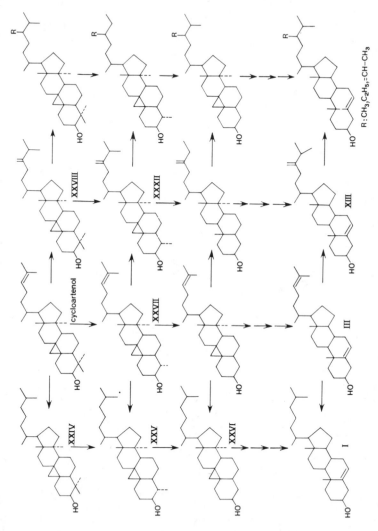

Fig. 18   Possible transformations, starting from cycloartenol (initial loss of the CH₃ group in position 4).

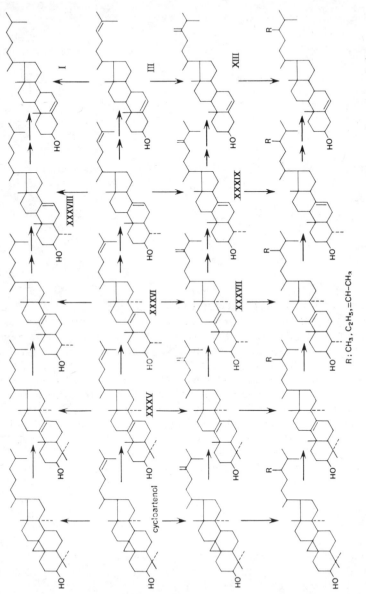

Fig. 19 Possible transformations starting from cycloartenol (early opening of the cyclopropane ring).

Biological function

The biological significance of C_{28} and C_{29} phytosterols is not wholly understood. However, their role in the constitution and function of cell walls seems established. Edwards and Green (1972) compared *in vitro* the ability of a variety of sterols to be incorporated in liposomal and erythrocyte membranes. Cholesterol is incorporated more readily than campesterol which also penetrates the membrane faster than β-sitosterol. It therefore seems that 24-methylation of these sterols reduces their incorporation in membrane phospholipids. A double bond in position 22 which reduces side-chain flexibility diminishes further the incorporation ability in phospholipids. Experiments performed by Grunwald (1971) showed that OH-free sterols present in membranes reduce the loss of electrolytes and other cell constituents (Bean, 1973).

C_{28} and C_{29} sterols can be precursors of various phytoecdysones (see above). In the alga *Achlya bisexualis* (and also in *A. ambisexualis*) a C_{29} steroid has been found which has hormonal activity: antheridiol. Antheridiol which is secreted by female gametes stimulates male antheridia formation and then induces secretion by the male gamete of a hormone which speeds up female sexual organ development (see later).

Unusual sterols of marine invertebrates

In a marine environment, and in particular in the case of invertebrates which are hosts to unicellular algae colonies, unusual methylations of side-chains leading to C_{29} and C_{30} sterols have been observed. A sterol with a propylidene group in position 24, 29-methyl isofucosterol **134**, has been isolated by Idler *et al.* (1970), from the mollusc *Placopecten magellanicus*. The biosynthetic pathway for this sterol suggests the possible existence of a 28(29)-methylenic precursor.

de Luca *et al.* (1972) established the structure of aplysterol **135** and 24(28)-dehydroaplysterol which have been isolated from the sponges *Aplysina* (*Verongia*) *aerophoba*. Gorgosterol **136** which is found in gorgonians results from three methylations of the side-chain which introduced two methyl groups in positions 23 and 24 and a cyclopropane in 22,23. An analogue, Δ_7 **137**, a 23-demethyl gorgosterol **138** and a curious seco-derivative (9,11-seco, 5-gorgostene 3,11-diol 9-one) **139** have also been found in gorgonians (Ling *et al.*, 1970; Enwall *et. al.*, 1972; Scheuer, 1973).

Various authors have isolated a new sterol **140** with 26 carbon atoms from marine invertebrates: Idler *et al.* (1970) were the first to establish its structure (24-dimethyl 5,22(*trans*) choladiene 3-olβ). The origin of this sterol, which is also found in a marine phytoplankton and in red algae, is not yet established and is an interesting problem. Certain marine animals such as tunicates transform it into a Δ_7 analogue or

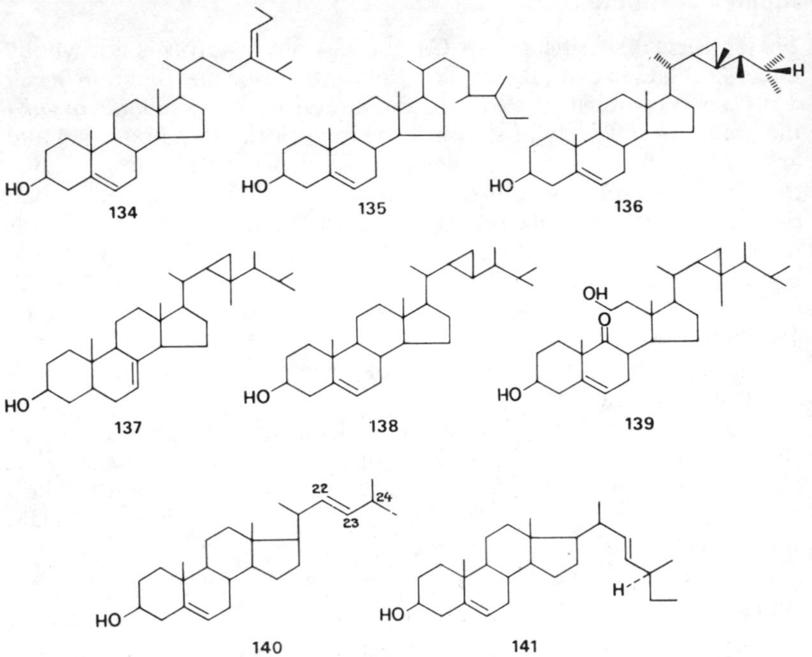

Fig. 20 Unusual sterols.

saturate the double bond at position 5 (Erdman & Thomson, 1972; Viala *et al.*, 1972; Kobayashi *et al.*, 1972). Kobayashi and Mitsuhashi (1974) isolated from an annelid the occelasterol **141** (which probably originates from a C_{28} sterol); The Δ_7 isomer (amuresterol) has been obtained by these same authors from the asteroid *Asterias amurensis*.

Chapter 4

Pheromones

Insect pheromones

Introduction

The word pheromone was first introduced by Karlson and Lüscher in 1959. The Greek roots *pherein* (to carry) and *hormon* (to excite) should have led to the word pherormone but, apparently for phonetic reasons, a letter was omitted. Although criticized by some, the word pheromone is now generally accepted. Karlson and Lüscher defined pheromones as substances secreted by an organism, often by a specialized gland, which, when perceived by an individual of the same species cause a specific reaction, a particular behaviour or exert some evolutionary influence. These long-distance effects imply either a vapour in air or a solute in water. At the present time pheromones causing a wide variety of effects are known, including sexual, trail-marking, aggregation-inducing, alarm and territory-marking (Karlson, 1960; Butler, 1967, 1970). The first pheromone to be isolated and identified was bombycol, a sexual pheromone secreted by the female moth *Bombyx mori* (Butenandt *et al.*, 1961).

Reproduction of the species is one of the main purposes of living organisms and many ingenious mechanisms have evolved to facilitate the process. An important aspect of reproduction is the attraction of members of the same species but of opposite sex. In order to achieve this it has been necessary to evolve very powerful and specific mechanisms. Powerful because the individuals may be far from one another, and specific because the individuals must be of the same species and of opposite sex. Many signals are used to induce sexual attraction, for instance changes in appearance (vision), emission of sound and of light. The naturalist Fabre was probably the first to study the existence of pheromones in insects. Fabre had shown the long-distance attraction which the female butterfly *Saturnia pyri* could exert on the male and had shown that this effect was due to an odour which could be transferred to an inert material such as cotton wool or cork. Fabre demonstrated that the antenna was the receptor for the signal, since males whose antennae had been removed became unable to find the female. The long-distance attraction (often several kilometres) is not the only remarkable effect of pheromones: they can also produce conditioning of the male to mating (in other words an aphrodisiac effect). The existence of sexual pheromones has been demonstrated in many insect species and already many have been isolated and identified.

However, while the knowledge of insect pheromones has progressed in the past few years, progress has not been so striking as far as the other invertebrate groups are concerned. The amount of information so far obtained is, however, considerable. Birch, in the introduction to his book *Pheromones* published in 1974, apologizes for not being able to cover the whole field in one volume of 495 pages!

The cells or specialized glands that secrete pheromones are of many different types (Percy & Wetherston, 1974). Variations in the activity of these glands, which depend on the physiological state of the female, have been observed frequently. Newborn female insects may not be attractive and after copulation they may lose their attractive power. Pheromone production can be linked to the extrusion of a specialized gland which is retracted into the abdomen after mating. Activity can also vary with time of day (i.e. position of the sun) or weather conditions. The implantation of microelectrodes into the antenna or the brain (deuterocerebrum) allows one to measure antennal currents and to record electroantennograms. This enables precise measurements of sexual pheromone activity and sensitivity to be made (Schneider, 1957). Insect antennae are usually equipped with perforated hairs (they may be single or multiple perforations) and innervated by simple or branching nerve endings. The antennae are therefore sophisticated organs specialized for the detection of molecules. Such a mechanism suggested to Williams that butterflies are 'marvellous flying machines dedicated to the service of love'.

Measurement of the specific activity of pure, isolated pheromones has given rise to some remarkable results. For bombycol, the minimum concentration required to give a 50 per cent positive reaction when a glass rod is dipped in a solution containing the pheromone and placed at a distance of 1 cm from the antennae of a male *Bombyx mori* is 10^{-12} μg/ml (i.e. about 2500 molecules per ml). Studies by Schneider *et al.* (1968) have shown that the threshold of the sensitivity of *Bombyx mori* antennae is reached when they are exposed to an airflow of 60 cm/s containing 10,000 molecules per ml for a few seconds. Calculations which take into account the possible number of receptors on the antennae suggest that a single molecule reaching a receptor would be enough to produce the effect in the animal. The maximum sensitivity to a sexual pheromone known at the present time is that of the cockroach, for which the threshold concentration is 10^{-14} μg/ml (Law & Regnier, 1971).

The biological action of pheromones seems to be related to the three-dimensional shape of the molecule and in particular to its π-electron cloud (Dravnieks, 1966). The importance of *cis* and *trans* isomers and of the distance between functional groups is also apparent. (This aspect will be examined later using two examples, bombycol and the queen substance of honey-bees.) The study of the physiological actions of pheromones became more complicated when it was realized that glands sometimes produce more than one pheromone which can act synergically; the significance of this at the level of chemoreception in the antennae is still unclear.

In some cases the effects of pheromones are delayed, for instance the royal jelly of the queen bee. This led Wilson (1965, 1968) to define two pheromone categories; *releasers,* with instantaneous effect (sexual pheromones) and *primers,* with a delayed effect.

In the early days of pheromone research it was believed that all were species-specific. However this is not true for all insects; Schneider (1962), Shorey *et al.* (1968) and Payne *et al.* (1973) have clearly shown that there is a lack of specificity in certain cases.

One of the techniques used in obtaining, purifying and identifying pheromones involves organic solvent extractions which are carried out either on whole insects or on isolated glands after dissection. This method is destructive inasmuch as it produces the amount of pheromone present in the gland at the time of death. A non-lethal technique can be used by passing air through a chamber in which a large number of females are bred and, by cooling the air to $-60\,°C$, condensing the substances thus carried away. Owing to the very small amounts of chemical obtained, the most sensitive analytical methods have to be employed: gas chromatography, preferably linked to mass spectrometry, spectrometry with Fourier transforms, micro-reaction analyses to determine functional groups, etc. The chemical structure of pheromones is often simple, yet the very small quantities obtainable mean that it takes a considerable time to determine their structure (e.g. 20 years for bombycol). This is probably what prompted a lady listener to ask during a lecture by Butenandt on the sexual pheromones of the female *Bombyx mori*: 'Oh, Dr Butenandt, why do you waste your time with butterflies?' (Butenandt, 1959).

Sex pheromones

Bombycol is the sex pheromone of the female butterfly *Bombyx mori* (mulberry tree *Bombyx*), and was the first pheromone to be purified and identified (Butenandt *et al.,* 1961). One million cocoons were used during the first purification procedure. They produced 300,000 females, but from all this material only 3 mg of a bombycol derivative was obtained. A second purification procedure was tried starting with 500,000 females which were dissected in order to obtain their glands and which produced 12 mg of bombycol derivative. Bombycol **142** consists of a hexadecadienol (Fig. 21) with a threshold activity of 10^{-12} μg/ml. The stereochemistry of the molecule is most important in terms of its biological activity. The natural conformation is 10-*trans*,12-*cis* and the activity of synthetic isomers is as follows:

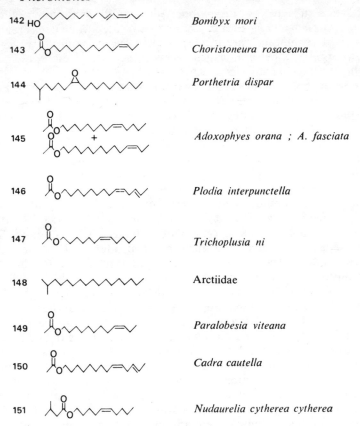

142 HO⌒⌒⌒⌒⌒⌒⌒⌒⌒   *Bombyx mori*

143   *Choristoneura rosaceana*

144   *Porthetria dispar*

145   +   *Adoxophyes orana ; A. fasciata*

146   *Plodia interpunctella*

147   *Trichoplusia ni*

148   Arctiidae

149   *Paralobesia viteana*

150   *Cadra cautella*

151   *Nudaurelia cytherea cytherea*

Fig. 21 Examples of sex pheromones in Lepidoptera; structures positively identified (Evans & Green, 1973).

10-*trans*,12-*trans*: 10 μg/ml; 10-*cis*,12-*cis*: 1 μg/ml; 10-*cis*,12-*trans*: 10^{-3} μg/ml (Butenandt *et al.*, 1962).

Most sex pheromones in the Lepidoptera for which the structural formula is known (Fig. 21) are alcohols or long-chain acids with 12–16 carbon atoms, some of which are esterified (acetates). Branching structures are exceptional. Interesting observations illustrating the complexity of the phenomenon were made on the pheromone of two lepidopterans of the family Tortricidae (*Argyrotaenia velutinana* and *Choristoneura rosaceana*), 11-*cis*-tetradecenyl acetate **143**. If one mixes this pheromone with synthetic isomers: 11-*trans*, 9-*trans*, 7-*trans* or 5-*trans*, etc., or if one replaces the acetyl group by a propionyl radical, an inhibition of activity is observed in the course of field captures. On the other hand, mixing the pheromone with saturated alcohols (dodecenyl alcohol for instance) increases the activity, in spite of the fact that these substances are inactive by themselves. These triàls emphasize the notion of synergistic action, as well as the equally important notion of inhibition. Such modifying effects on the pheromone action, of course, favour practical applications (Roelofs & Comeau, 1971). In the same Tortricidae family, 'the distorter of spruce buds', *Choristoneura fumiferana* produces 11-*trans* tetradecen 1-al as a sex pheromone, and the corresponding tetradecanol, which acts as an antagonist (Weatherston & MacLean, 1974). A similar thing was reported by Cardé *et al.* (1973) with *Porthetria dispar*. The sex pheromone of this butterfly (Bierl *et al.*, 1970) is the 7,8-*cis* epoxy 2-methyl octadecane; 2-methyl 7-octadecane, perhaps its biological precursor, is an antagonist of its action. The existence of substances with synergistic and antagonistic actions appears to be quite widespread among the Lepidoptera (Roelofs & Cardé, 1974). In the case of *Archips semiferanus*, a lepidopteran whose caterpillars, the oak leaf rollers, ravage forests in the North-East of the USA, sexual attraction is brought about by a complex mixture of pheromones consisting of a series of diversely unsaturated tetradecenyl acetates (Hendry *et al.*, 1975). Kasang (1973) demonstrated the enzymatic inactivation of the pheromone fixed on the antenna, using ³H-bombycol. (For further reviews see Eiter, 1970; Karlson & Schneider, 1973; Evans & Green, 1973; Jacobson, 1966).

In 1969 Butler suggested the name 'aphrodisiac' be applied to any chemical released either by the male or by the female which prepared the mate of the opposite sex for intercourse after formation of the pair, if this were brought about by the effect of pheromones or by some other means (Birch, 1974).

Pheromones can indeed have two sorts of effects, they can attract members of the opposite sex and they can be aphrodisiacs. In the *Danais*-type butterflies, the male possesses, at the tip of his abdomen, a tuft of scented hair (a hair pencil) which is opened up when the female approaches and whose emanations promote copulation. From this organ (in *D. gilippus, D. chrysippus, Lycorea ceres*) Meinwald *et al.* (1969, 1971) isolated pyrrolizidinone **152**. This ketone inhibits the flight of females, which immediately land and allow themselves to be readily approached by males. Males are attracted by plants of various families containing the alkaloids which are precursors of this pheromone: *Heliotropium, Tournefortia, Cynoglossum, Senecio, Eupatorium, Crotalaria,* etc. (Edgar & Culvenor, 1975).

Certain substances found in the sperm of insects induce physiological and behavioural reactions in the females, such as acceleration of oocyte maturation and of egg-laying and refusal of subsequent copulation. This monocoitic behaviour is caused in *Drosophila funebris* by a 27 amino-acid peptide (Baumann, 1974; Baumann *et al.*, 1975). The active products fit the definition of pheromones; however, one often encounters in the literature the words 'matrones' or 'paragonial substances' which emphasize the origin of these substances: the paragonial glands of the male. The male of the cockroach *Blatella germanica* displays a characteristic behaviour

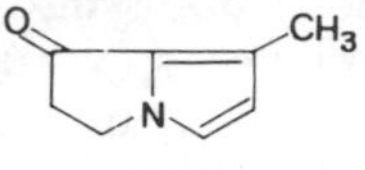
152

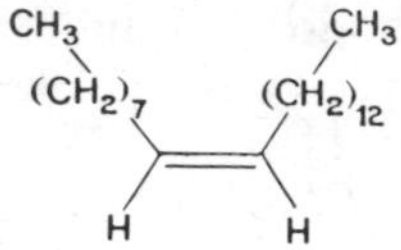

153

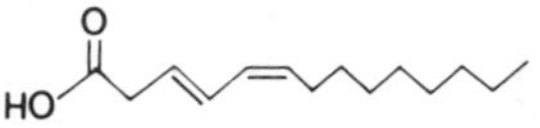

154

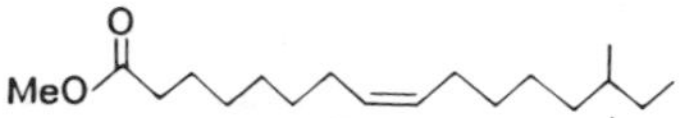

155

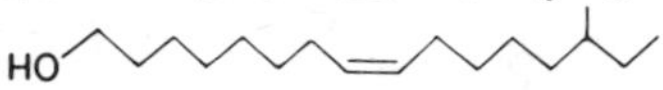

156

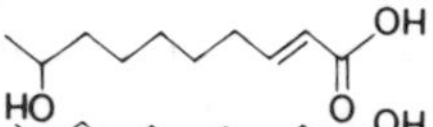

157

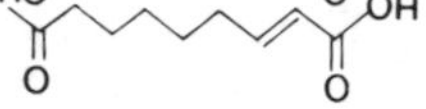

158

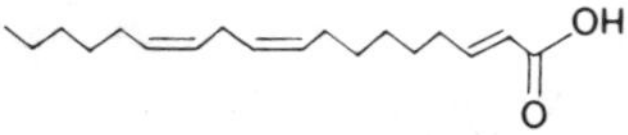

159

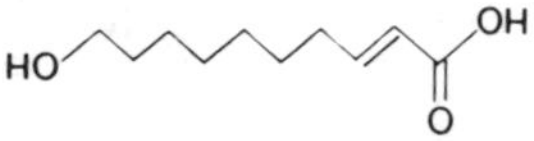

160

Fig. 22 Various pheromones of insects.

in the presence of the female. Following contact with her antennae, he becomes agitated and lifts up his wings and this characteristic position precedes mating. Nishida *et al.* (1974) isolated two substances from the wax which covers the body of the females; one of these, which was obtained in crystalline form, was identified as 3,11-dimethyl 2-nonacosanone and can be considered as an aphrodisiac. In comparison with Lepidoptera, few pheromones have been identified in other insects. Muscalure **153,** the pheromone of the house-fly, has a rather low attractive potency. The coleopteran *Attagenus megatoma* utilizes the unsaturated fatty acid **154**, *Trogoderma inclusum* the pheromones **155** and **156** (Rodin *et al.,* 1969; Silverstein *et al.,* 1967).

Simpler molecules can sometimes be the active principles, such as valeric acid (*Limonius californicus*) and phenol (*Costalytra zeelandica* (Jacobson *et al.*, 1968; Henzell & Lowe, 1970). 2-*Trans* hexen 1-ol acetate has been found by Butenandt and Tam (1957) in the male aquatic beetle *Belostoma indica* (*Lethocerus indicus*). This substance is responsible for the scent liberated by the insect and is used as seasoning in certain parts of Asia. Its biological role remains unclear, but since it is only secreted by the male it might well be an aphrodisiac.

A pheromone with multiple activities: the queen substance of the honey-bee

The mandibular glands of the queen honey-bee (*Apis mellifera*) secrete biologically active substances which have been studied by Callow and Butler and by Barbier and Pain. The activities of this secretion can be summarized as follows:

1. Attraction of worker bees toward their queen
2. Attracted worker bees lick the body of their queen thereby swallowing active substances, among which is the queen substance (9-oxo-2-*trans* decenoic acid **158**). This results in the blocking of their ovaries (only the queen bee can lay)
3. The queen substance inhibits the construction of royal cells; it therefore has an informative role since it signals the uselessness of raising another queen
4. The queen substance contributes (together with its reduced form, 9-hydroxy-2-decenoic acid **157**) to the attraction of males towards the queen, during her rare outings.

The queen substance is therefore a social pheromone, a sex pheromone and at the same time an anti-hormone (Callow & Johnston, 1960; Pain & Barbier, 1963; Pain *et al.*, 1962; Butler, 1967; Barbier, 1968; Barbier & Lederer, 1960). Synergistic effects are probably very important because the queen substance is never as active in the course of biological assays as it is when tested in the presence of other mandibular constituents. In the case of the attraction of worker bees toward the queen, it is the acid mixture (including the queen substance) which is active – the chromatographically separated components are never as active as the reconstituted mixture (Barbier *et al.*, 1960; Pain *et al.*, 1962). This attraction is essential for the normal functioning of the hive since, after licking the body of the queen, worker bees will transmit from mouth to mouth in the colony the pheromones thus collected (transmission in the course of food exchanges: the phenomenon of trophallaxis).

The queen substance also differs from other substances in its activity threshold; the inhibition of the construction of royal cells requires 100 μg per day for a population of 200 bees. It is true that this substance must be ingested and that its antennary perception seems to be a secondary phenomenon. The mandibular glands of the queen honey-bee can contain large quantities of queen substance, up to 1 mg, depending on the season.

The biological activity of the queen substance is closely related to its structure. Any variation, even if it only involves the suppression of one carbon atom between the methyl-ketone and the unsaturated α-β acid, abolishes the activity. Such a result suggests, even in the case of the inhibition of royal cell construction, the presence of a specific receptor with a double site (Barbier & Hügel, 1961; Pain *et al.*, 1962; Barbier, 1968).

An analogue of the queen substance, 10-hydroxy-2-*trans* decenoic acid **160** is found in large quantities in the substance produced by the mandibular glands of worker bees. The biological significance of this substance is not well understood but it certainly contributes to larval development since royal jelly constitutes, together with pollen, the usual diet of the larvae.

The biosynthesis of queen substance (and that of the acid of the royal jelly) has still to be elucidated. Hopkins *et al.* (1969) put forward a hypothesis according to which the queen substance might be metabolically produced by the honey-bee from an unsaturated fatty acid **159** normally present in certain pollens (such as that of clover), since this acid has the property of attracting honey-bees toward the pollen.

Since the queen substance has both an instantaneous effect, when it is a sexual pheromone, and a delayed action when its role is that of a social pheromone, it is either a releaser or a primer according to the definitions of Wilson (1962, 1965). Queen pheromones act together with substances released by the Nassanoff gland of worker bees (Morse & Boch, 1971) to ensure the cohesiveness of the swarm (the pheromones of this gland are geraniol (Boch & Shearer, 1962a, 1962b), nerolic and geranic acids and citral (Weaver *et al.*, 1964; Butler & Calam, 1969)).

2,3-Dihydro-farnesol (*trans*), its acetate and ethyl laurate are found in the mandibular glands of the male bumble bee *Bombus terrestris*. These are used for territorial markings and attract both males and females (Stein, 1963; Bergström *et al.*, 1968).

δ-n-Hexadecanolactone is a pheromone which stimulates the construction of royal cells; it is produced by the queen of the hornet *Vespa orientalis* (Ikan *et al.*, 1969).

Alarm pheromones

The concert of pheromones among insects which are living in societies or are momentarily grouped is a subtle one. Multicomponent systems are generally used, and these can originate in more than one gland. The variation in the meaning of messages depending upon the amount of substance released is a probable mechanism, but little is known about this complicated aspect of intraspecific interactions. For example, the ant *Acanthomyops claviger* possesses two alarm pheromone reservoirs. The mandibular glands secrete the terpene series **166** to **170** (among which are citronellal **168**, nerol **169** and geranial **170**). The Dufour gland in the abdomen contains three hydrocarbons **161**, **162** and **164** as well as two methylketones **163** and **165** (Table 7). All these substances are efficient alarm pheromones, with the exception of pentadecane **164** and 2-pentadecanone **165** (Regnier & Wilson, 1968). When an ant of this species is in distress it releases small amounts of

Table 7 Examples of alarm pheromones produced by Hymenoptera (references cited or reviewed by Blum, 1974)

Substances	Insects
2-heptanone	*Iridomyrmex* (Formicidae); *Trigona* (Apidae)
4-methyl-2-hexanone	*Dolichoderus* (Formicidae)
4-methyl-3-hexanone	*Manica* (Formicidae)
6-methyl-5-heptene 2-one	*Tapinoma* (Formicidae)
3-octanone	*Myrmica* (Formicidae)
4-methyl-3-heptanone	*Pogonomyrmex* (Formicidae)
2-methyl-4-heptanone	*Tapinoma* (Formicidae)
2-nonanone	*Trigona* (Apidae)
3-nonanone	*Myrmica* (Formicidae)
6-methyl-3-octanone	*Crematogaster* (Formicidae)
3-decanone	*Manica* (Formicidae)
4, 6-dimethyl-4-octene-3-one	*Manica* (Formicidae)
2-hexene-1-al	*Crematogaster* (Formicidae)
citronellal	*Acanthomyops* (Formicidae)
citral	*Acanthomyops* (Formicidae); *Trigona* (Apidae)
formic acid	*Formica* (Formicidae)
n-decane	*Camponotus* (Formicidae)
n-undecane	*Acanthomyops* (Formicidae)
n-dodecane	*Camponotus* (Formicidae)
dimethylsulphide	*Paltothyreus* (Formicidae)
dimethyltrisulphide	*idem.*

Alarm pheromones in *Acanthomyops claviger*

Dufour gland   Mandibular glands

alarm pheromones which quickly warn its neighbours. Attracted by these substances they gather towards the ant responsible and release more pheromones themselves. By such a mechanism, the whole ant-nest is quickly brought to a state of emergency. Alarm pheromones are widespread among insects (Maschwitz, 1964; Blum, 1969). Table 7 illustrates the wide range of molecules encountered (Blum, 1974; see also Bergström & Löfqvist, 1970, 1971, 1972). Bergström and Löfqvist (1973) analysed the contents of Dufour glands in *Formica nigricans, F. rufa* and *F. polyctena*. Forty-six volatile substances were found, with little difference occurring between the three species. Linear hydrocarbons predominate, with undecane **161** as the main component. Unsaturated and branching hydrocarbons are present in lesser quantities. Among the less volatile substances the authors identified all-*trans*-geranylgeraniol and its acetate, octadecenyl acetate as well as the acetates of other aliphatic alcohols (Bergström & Löfqvist, 1972).

When bees sting an enemy, as well as injecting venom, they also inject a mixture of isoamyl acetate, propionate and butyrate, which will prompt other bees to come and sting the same victim (Boch *et al.*, 1962; Gunnison, 1966). This phenomenon, which had already been observed by Huber in 1814, is well known by apiculturists, who remark upon the banana scent of bee stings (Gary, 1974). A ketone, 2-heptanone, which is present in the mandibular glands of worker bees is thought to be an alarm pheromone with a different mode of action (Morse *et al.*, 1967).

Trans β-farnesene is an alarm pheromone isolated from three aphid species (Bowers *et al.*, 1972). In the pyrrhocoride *Dysdercus intermedius* the active product is 2-hexenal, as in the case of the ant *Crematogaster africana* (Calam & Youdeowei, 1968; Bevans *et al.*, 1961; Blum *et al.*, 1969).

The use of alarm pheromones has been observed in many termite species (Moore, 1974). (+) Limonene has been found in Australian termites *Drepanotermes rubriceps* and *D. perniger*. Terpinolene is secreted by *Amitermes herbertensis*. Another terpene, α-pinene, is used by *Nasutitermes exitiosus* (Moore, 1968).

The *Azteca* ants have in their anal glands a series of methyl cyclopentanones which have a defensive role, and which are also used as alarm pheromones (Wheeler *et al.*, 1975).

Trail pheromones

Trail pheromones are part of a system of scents whose function is the labelling of food or of trails leading to that food. As before, in the case of alarm pheromones, the meaning of these signals is still beyond our understanding. The identified substances are often only the main products of mixtures and their activity might be only part of the story. The leaf-cutting ant *Atta texana* marks its trail with a pyrrolic ester **171** (Tumlinson *et al.*, 1972). The termite *Zootermopsis nevadensis* uses caproic acid (Hummel & Karlson, 1968). *Reticulitermes virginicus* excretes a tri-unsaturated alcohol **172** (Matsumura *et al.*, 1968; Tai *et al.*, 1969). This acid has also been isolated from fungi-attacked wood on which the termites usually feed (see Carter *et al.*, 1972). 3-*Cis*-hexen-1-ol **173** (leaf alcohol) greatly attracts the termites *Calotermes flavicollis* and *Microcerotermes edentatus* from which it has been isolated. This alcohol probably has a vegetable origin

(Verron & Barbier, 1962). In the case of termites, marking by non-specific substances from the food is possible. These insects are also attracted by substances formed in decaying wood, which they appear able to accumulate and later use as trail-following markers (Becker, 1964).

Nasutene (neocembrene-A), a 14-membered cyclic hydrocarbon **174**, is the trail-following pheromone of a large group of Australian *Nasutitermes*, including *N. exitiosus* (Birch *et al.*, 1972). This hydrocarbon is found in large quantities in the Indian cedar *Commiphora mukul* from which incense is made (Moore, 1974).

Trigona subterranea labels the source of newly discovered food with a secretion of the mandibular glands which contains two citral isomers, neral and geranial, with the result that it attracts other workers (Blum, 1974). *T. spinipes* behaves in the same way, but with heptan-2-ol (Blum, 1974). The Dufour glands of solitary bees (*Halictus calceatus, Colletes cunicularis, Andrena*, etc.) contain macrocyclic lactones – hexadecanolide, octadecanolide, eicosanolide, etc. – which may be important for nest marking (Andersson *et al.*, 1967; Bergström, 1973).

Trail-following pheromones of ants can be detected by carnivorous insects, which then follow the ant-trail to feed on the nests, for instance the myrmecophilic beetles of the families Hysteridae, Staphylinidae and Limulodidae (Akre & Rettenmeyer, 1968). Mites, myriapods and even snakes are thought to have also discovered this convenient way of detecting ant nests (Watkins *et al.*, 1967).

Aggregation pheromones

The sex pheromones of the Coleoptera are difficult to distinguish from aggregation pheromones, their actions often being identical. In limited cases their activities are traceable to very simple molecules: valeric acid, phenol, etc. In most cases, however, they are due to several compounds acting synergistically. This is the case, for instance, in a beetle which lives on tree-bark, *Ips confusus*. In the course of boring bark, these insects throw out a mixture of wood-dust and secretions which have an effect on both the males and females. Three terpenic alcohols have been identified, all of them being a prerequisite of biological activity (see **175**, **176**, **177**). From the secretions of another beetle, *Dendroctonus brevicomis*, a compound with a curious structure has been isolated, bicyclic dioxane **178**, which is called exo-brevicomine. This compound attracts both sexes and myrcene exerts a synergistic action. In a closely related species, *Dendroctonus frontalis,* the activity is due to a mixture of verbenol **177** and of another dioxobicyclic derivative, frontaline **179** (Renwick & Vitré, 1970; Evans & Green, 1973). Frontaline is found in newborn females of *D. frontalis* and thus might not derive from food; its attraction over males is manifestly stronger than over females. It should be noted that other

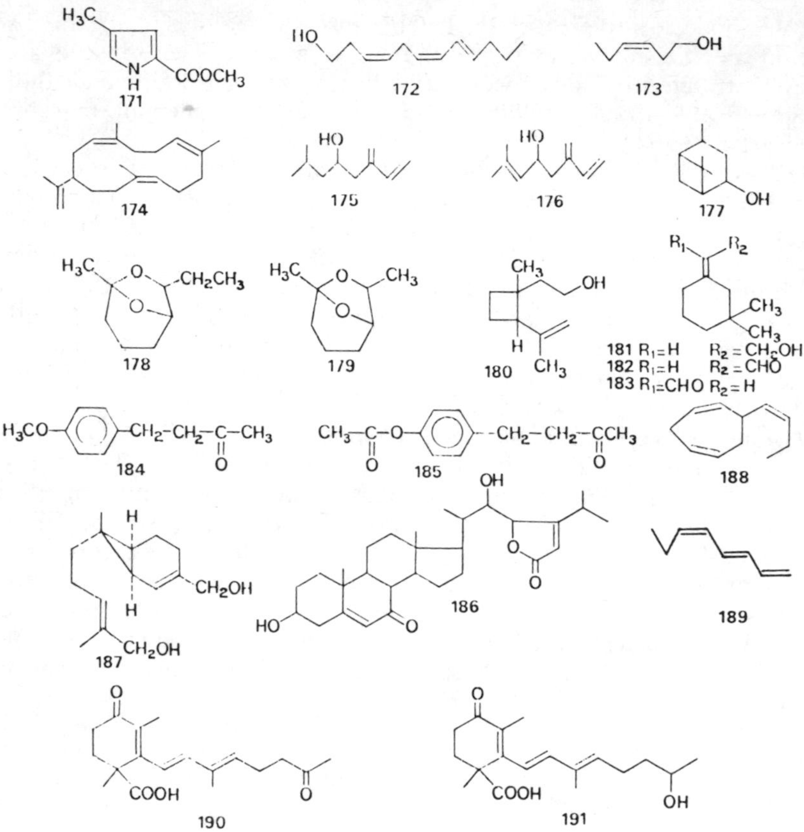

Fig. 23   Various pheromones.

terpenes have synergistic properties, one example being 3-carene (in *D. brevicomis*), although its potency is not as great as that of myrcene. α-Pinene and camphene are two other synergistic compounds found among *Dendroctonus* (reviews by Silverstein, 1970; Borden, 1974; Lanier & Burkholder, 1974).

In one instance, it has been possible to establish that active products are synthesized by the intestinal flora during digestion. *Bacillus cereus,* isolated from the male or female beetle *Ips paraconfusus,* is able to transform α-pinene into verbenol. Other bacteria from different *Ips* species can bring about this transformation, which suggests a possible origin for the pheromone: a chemical transformation of terpenes found in foodstuffs (Brand *et al.,* 1975).

A mixture of the ethyl esters of palmitic, linoleic, oleic and stearic acids with the methyl ester of oleic acid constitutes the scent responsible for the aggregation of the coleopteran *Trogoderma granarium* (granary beetle) (Yinon *et al.,* 1971).

The male of the boll weevil (*Anthonomus grandis*) is attracted by the terpenes in the buds of the cotton plant ((+)α-pinene, (−)limonene, (−)β-caryophyllene, (+)β-bisabolol and caryophyllene oxide). After feeding, the weevils excrete an active mixture of four new terpenes **180, 181, 182** and **183** which act synergistically (Tumlinson *et al.,* 1969).

The aggregation of beetles of the family Scolytidae is essentially a social phenomenon whose first result is the formation of mixed-sex groups. Male–female attraction with its sequel, mating, then follows. The entire phenomenon is dependent upon the insect's relationship with the host-plant. When the insect has found a good supply adequate for its species, the mechanism leading to reproduction is triggered, starting with the release of a biologically active complex in the faeces.

The use of attractants in insect control

Sex pheromones have been used with success in trapping insect pests so as to minimize the use of pesticides. This additional factor in crop control enables the question of intervention to be more accurately assessed both in terms of time and dosage (Bethel *et al.,* 1972; Batiste, 1972). In areas of forests a method such as this allows large areas to be brought rapidly under control. The setting of traps in order to attract predators with pheromones have led to interesting results (Jacobson, 1972). In attempts to eradicate insects, it might well be questioned, however, whether the number of insects captured could warrant a complete rejection of the conventional methods for using insecticides. Sight should not be lost of the fact that saturation of the atmosphere with pheromones, resulting in a constant presence of the signal, makes detection singularly difficult. Habituation of the nervous system of males to a pheromone level considerably higher than its threshold can cause inhibition (Bartell & Lawrence, 1973; Tette, 1974). Aggregation pheromones have been tried in forests on one or, in some cases, a few selected trees which, after fixation of the populations, are felled (Knopf & Pitman, 1972). In fact, although there were great expectations following the discovery of pheromones, it has to be admitted that their use so far has not reached that degree of eradication likely to justify discontinuing insecticide use. Looking more to the future, however, a carefully balanced combination of the two methods is a reasonable proposition.

After considerable efforts in the field of synthesis, a series of very active molecules has been found. Previously it had been demonstrated that eugenol strongly attracts the Japanese beetle *Popillia japonica*, in spite of the fact that it is not a pheromone (Yoonwardede *et al.,* 1970). Trimedlure, a mixture of four chlorinated isomers, effectively lures the fly *Ceratitis capitata* (Beroza, 1970). Bartell *et al.* (1957), had already established the attracting powers of anisylacetone **184** on the male of the Hawaiian melon fly *Dacus cucurbitae*. Beroza *et al.* (1960)

synthesized   an   even   more   active   analogue, p-acetoxy-
phenyl-2-butanone **185**. These compounds have been used in the
control of *Dacus*.

Attraction by non-naturally occurring substances is an interesting
phenomenon, more especially since their effect is the same as that of
real pheromones, the males only of specific species being attracted.
Djerassi *et al.* (1974) published a detailed analysis of the possibilities
of insect pest control. The cost of producing active substances is the
determining factor in the feasibility of control by these means. Except
in the case of considerable economic advantage, industry is likely to be
very reluctant to start large-scale production of these chemicals. At the
moment, industries focus their research on the improvement of
conventional insecticides, chlorinated insecticides, organophosphates
or carbamates. It is the lack of specificity of these insecticides which is a
serious drawback for some, and a major advantage for others. Djerassi
et al., suggest the term 'biorational agent' for any active material, the
use of which is specific. Pheromones, hormones, anti-hormones all fall
within this definition. So does the use of microbial insecticides. Two
micro-organisms are commercially available in the USA at the
moment: *Bacillus thuringensis* and *Bacillus popillae*. 'Biological
control' consists in the use of predators or parasites which are selective
for the species to be eliminated. Eighty programmes of 'biological
control' are known to have been successful. Some figures might be
quite useful here in showing the importance of the financial side of
insecticide production. The development of a new product costs in the
order of 11 million US dollars and the time for development of the
product can be as long as ten years. The development of hormones and
anti-hormones is still at the experimental stage. Pheromones have
been used on a local scale. The sexual pheromone of the house-fly
(muscalure) is on sale in the USA in a mixture of conventional
insecticides; the attractive power of this pheromone is known to be
weak but nevertheless the presence of muscalure increases its power
significantly. Disparlure, the sexual pheromone of the predatory forest
butterfly *Porthetria dispar,* and grandlure, the sexual pheromone of
the predatory cotton plant beetle *Anthonomus grandis*, will soon be
commercially available too. In spite of the fact that the problems
involved in the synthesis of biorational agents on an industrial scale are
just as complex as those for insecticide production, the former have the
advantage of being active at considerably greater dilutions than
conventional insecticides. Djerassi *et al.* (1974) consider that the
problems of controlling insects are of equal importance to those
associated with human birth control.

Pheromones in algae and fungi

The term pheromone seems to be reserved for higher organisms and is

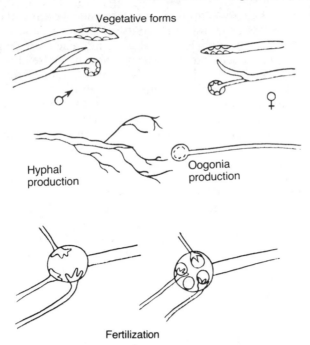

Fig. 24 Various stages in the reproduction cycle of *Achlya bisexualis.*

not generally used to describe compounds responsible for intraspecific interactions in the plant kingdom. This distinction seems even more subtle since English-speaking authors do not hesitate to use the word hormone to describe substances responsible for sexual attraction in algae and lower fungi. This extra use of the word hormone leads to the loss of an important factor; the internal aspect, and extends the definition to 'a substance produced by an organism, and transported in some way to another part of the same organism *or to other organisms of the same species* where it results in specific responses'. Considering what has been said above, such a definition could very well apply to pheromones for insects. German literature uses the term *'gamone'* which is undoubtedly better suited since these substances are responsible for the attraction between gametes (Reschke, 1969; Raper, 1952; Jaenicke & Müller, 1973). In the slime mould *Achlya bisexualis*, hyphae are sterile but react either as males or females, depending on their neighbour. If a male hypha is close to a female one, ramifications will form among the male hyphae, leading to antheridia. Enlarged branches develop on the female hyphae, with a characteristic dense and globular structure: these are the oogonia. Antheridial growth will continue towards the oogonia, with which they will eventually join (Fig. 24). As soon as contact has been established, the end of the antheridium will differentiate into a gametangium, or antheridium and eggs will appear in the oogonia. Towards the end of this process, tubes grow into the cavity of the oogonium from the antheridium and a male nucleus is released into each female gamete. The whole of this developmental process depends on 'sex hormones'. The activation of male hyphae, control of their growth and transformation are under the influence of antheridiol (Arsenault *et al.*, 1968). Antheridiol has a steroid structure (186) and its biosynthesis consists in the transformation of a very common phytosterol among fungi, fucosterol (XIV, Table 4); (Popplestone & Unrau, 1974; Barksdale, 1969). Sirenin 187, which is produced by *Allomyces* (Phycomycetes) was the first sex hormone to be isolated from plants. In these fungi, the secretion of this compound from the female gametes results in positive chemotaxis among the male gametes. Its structure was determined by Machlis *et al.*, (1966, 1968) and a possible biosynthetic pathway has been postulated using *cis*-farnesyl pyrophosphate as a starting material. In cultured *Chlamydomonas*, a unicellular green alga, clumping has been observed, out of which copulating cell pairs appear. The substances responsible for these attractions might be glycoproteins excreted by flagella (Förster & Wiese, 1954; Wiese & Jones, 1963).

In the brown alga *Ectocarpus silicosus* gamete attraction is brought about by ectocarpene 188 which is produced by the female gamete. This hydrocarbon with a 7-membered ring might be genetically linked to dictyopterenes (Müller *et al.*, 1971; Jaenicke & Müller, 1973). Spermatozoids of the brown alga *Fucus serratus* are attracted towards the eggs by fucoserraten 189 (1,3,5-octatriene) (Müller & Jaenicke, 1973). In another brown alga, *Cutleria multifida*, Jaenicke *et al.* (1974) have shown the presence of multifidene and aucantene, additional hydrocarbons of the gynogametes, which attract the male gametes.

In the heterothallic fungi of the Mucorales type, sexual development is controlled by trisporic acids 190 and 191 (Austin *et al.*, 1969). By mixing cultures of the (+) and (−) forms of *Blakeslea trispora*, the biosynthesis of these acids from β-carotene is de-repressed in both partners (Bu'lock, 1973). The precursors in the biosynthesis have been identified (prohormones); only the (+) sexual mycelium is able to transform the prohormone produced by the (−) form into trisporic acid, and vice versa (Bu'lock *et al.*, 1974).

Vertebrate pheromones

The production and release of scents by specialized glands in mammals is a widespread phenomenon. These glands are often found close to the sexual organs and their activity varies according to sexual activity. Their role is, however, less clearly understood than in insects, and their numerous effects, for instance, sexual attraction, species identification and territorial marking, often cannot be distinguished from each other in any clear way. These scents are produced from complex chemical mixtures and we may refer to them as 'perfumes'. The concept of pheromone, as defined earlier, becomes confusing and the specific effects of these compounds are difficult to distinguish from Pavlovian phenomena. Indeed, since the most common effect consists in the male locating a scent which is characteristic of the receptive female, the distinction between innate behaviour and that which results from learning becomes important.

In 1950 Lederer published a review of scents released by animals. Little progress has been made since then. Among identified compounds we may note muscone **192**, the substance released from the scent gland of the male ibex *Moschus moschiferus,* and civettone **193** released by the civet cat *Viverra zibetha.* In the male and the female rat *Ondatta zibethicus rivalicius*, both exaltone **194** and civettone are found. All these chemicals have a common characteristic: they are all large cyclic ketones and are all of interest to the perfume-maker because of their musky scent. Similar compounds are found in the duck *Ana moschata*, in the tortoise *Cinosternon pennsylvanicum,* in certain alligators, etc. In the latter, four pairs of scent glands are found which produce yaccarol, a mixture containing citronellol. In the black-tailed deer *Odocoileus hemionus colombanus*, a pheromone responsible for group behaviour has been identified (a 12-carbon hydroxy-acid, also stable in the lactone form; **195**) (Brownlee *et al.,* 1969; Müller-Schwartze, 1969; Ralls, 1971).

The subauricular gland of the antelope *Antilocapra americana* (male pronghorn) releases a scent which consists of a mixture of valeric and

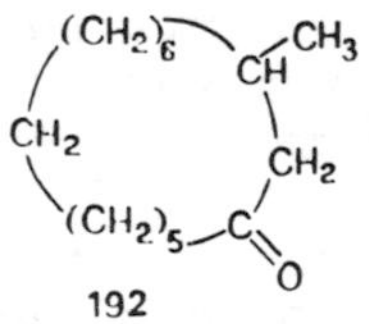

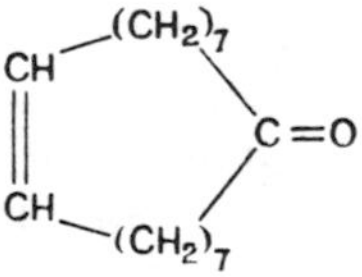

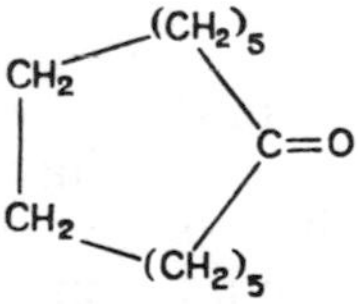

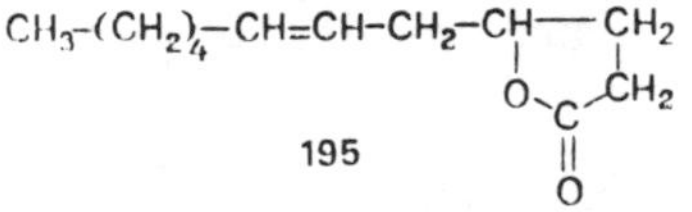

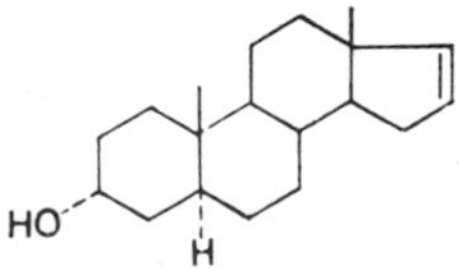

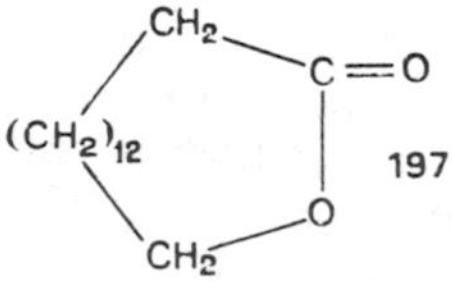

Fig. 25 Vertebrate pheromones.

isovaleric acids and esters with branching aliphatic alcohols (Müller-Schwartze, 1974). The Mongolian gerbil *Meriones unguiculatus* uses a ventral gland for territorial marking. Its active substance is phenylacetic acid (Thiessen *et al.*, 1974).

Mustelidae secrete foul-smelling mixtures containing mercaptans and sulphides from their anal glands. The Philippine mydaus (*Mydaus marchei*) can eject its secretion more than 100 cm. A similar mechanism is found in the South-American skunk *Conepatus suffocans* and in its North-American counterpart, *Mephitis mephitis nigra*. The latter can throw an evil-smelling liquid onto its enemies at a distance of 300 cm and the smell can be detected several hundred yards away. Dicrotyl sulphide has been isolated from this powerful defensive weapon (review by Lederer, 1950). Vertebrates possessing such deterrents often have black and white speckled coats which remind one of the aposematic characteristics encountered among insects. The glutton (wolverine) *Gulo luscus* found in Arctic regions secretes in its urine a foul-smelling substance secreted by specialized glands, which might act as a territorial marker. The smell is so strong that houses which this animal manages to enter become uninhabitable unless thoroughly cleaned.

Wheeler *et al.* (1975) isolated from the anal glands of the hyena 5-thiomethyl-pentane-2,3-dione; this chemical is the first naturally occurring α-diketonic thioether to be identified.

In the male pig, maxillary glands produce a smelly substance which acts on receptive females; the response is of a behavioural nature (inhibition of movement which facilitates copulation). This immobilizing response is due to a steroid, 5α-16-androstene-3α-ol **196**; (Patterson, 1968; Melrose *et al.*, 1971; Gower, 1972).

The study of mice pheromones has led to several interesting observations. The preputial glands secrete substances which are able to induce oestrus in groups of females, at a distance. Observed effects are rather complex, but can be summarized as follows:

Bruce effect: If a male from a different colony (with a different scent) is introduced into the cage of a newly fertilized female, egg rejection and a quick return to oestrus follows.

Lee-Boot effect: When a group of female mice (more than four) are kept together in the absence of a male, oestrus is suppressed and phantom pregnancies are observed in 61 per cent of individuals.

Ropartz effect: If a mouse colony is large enough, hypertrophy of the adrenal glands is observed in all individuals, leading to an increase in the corticosteroid levels and a decrease in the fertility of the colony.

Whitten effect: A smelly substance in the male urine induces and accelerates oestrus in the females. This effect can therefore be used to synchronize oestrus in a group of females by introducing a male, if oestrus had previously been suppressed by grouping the females (Lee-Boot effect) (Bruce, 1966; Whitten, 1966; Ropartz, 1966, 1968).

In the fish *Lebistes reticulatus* female oestrogens attract and cause hyperactivity in the males (Amouriq, 1965). Alarm pheromones have been demonstrated in fish. As far as we know, these substances, probably pterines, have not yet been identified. Isoxanthopterine induces flight in fish of the carp family. Skin extracts induce responses in the *Bufo* toads; bufotoxin, found in the venom, induces the same behaviour as whole skin extracts and might be an alarm pheromone (Pfeiffer, 1974).

Michael *et al.*, (1967, 1970) trained monkeys to depress a lever a certain number of times in order to gain access to the female. Using this technique it was possible to establish the relationship between the secretion of active substances by females and male motivation. The active scent consists of a mixture of various acids: acetic, propionic, isobutyric, butyric, isovaleric, etc. The same series of fatty acids had been observed in human vaginal secretions (Michael *et al.*, 1974). This study, carried out on 682 subjects, has demonstrated a cyclical variation in the amount of fatty acids present. They increase during the follicular stage and decrease during the high progesterone phase. Women using steroid contraceptives do not show this phenomenon.

It is likely that sex pheromones played a role in the life of primitive man. From this point of view, modern man is a degenerate animal for whom scent does not apparently play the role for which it was evolved. The role of vision, on the other hand, became more important as higher brain functions developed. In the female, the maximal olfactory sensitivity coincides with the peak of the follicular activity. Exaltolide **197**, a synthetic lactone, is sensed much more by women than by men and the threshold perception levels of smell vary with the ovarian cycle. Men who have been injected with low oestrogen doses develop the ability to detect exaltolide (the experiments of Le Magnen). Relations between nose–olfactory bulb–hypothalamus–sexual organs seem to be clearly established among vertebrates. The congenital absence of the olfactory bulb is always accompanied by genital underdevelopment for both sexes (Morsier syndrome). A difference in olfactory sensitivity between men and women and also between two women of different physiological states has also been observed with the scented steroid **196**, which has been mentioned previously. This steroid, a sex pheromone of the boar, is found in men's sweat. Brooksbank *et al.* (1974) suggested that in man it might play a role similar to that in pigs. All these various aspects, together with their psychological repercussions have been reviewed by Comfort (1971); see also Whitten and Champlin (1973).

McClintock (1971) has shown that hormonal regulation is affected by the presence of other individuals. This author discovered the synchronization of the ovarian cycles of girls living in situations of close contact such as boarding schools.

Chapter 5

Stability and instability: two examples

In this last chapter, two very different types of interaction have been brought together to emphasize their contrast. The interaction between plants and insects follows directly from the requirements of adaptation. Combinations of attractions and repulsions and the superposition of allowed and prohibited behaviour lead to a system of complex equilibria.

Rhythms condition life. Periodicities which are linked to diurnal or seasonal variations modulate the ecosystems, and trophic influences are related to these regular fluctuations. The notion of spatial coincidence (of which parasitism is the most extreme example) or the notion of temporal coincidence, play determining roles. A shift of one week in the appearance of a plant species can become a limiting factor in the development of insect species that rely on it. The need for spatial coincidence between the host and the consumer can sometimes lead to evolutionary change. In the case of the lepidopteran *Grapholita molesta* for instance, the first generation larvae feed on young Rosaceae shoots, whereas the following generation will live off the fruits (Labeyrie & Huignard, 1973). The nutritive plant which is used as the breeding ground is often at the same time the location for the meeting of the two sexes. In the case of *Antherea polyphemus* (Riddiford & Williams, 1967) and *Acrolepia assectella* (Rahn, 1968), the plant actually speeds up the production of sex pheromones by the females. The complexity of the relationship between plant and insect can readily be seen, as well as its usefulness in terms of the preservation of the animal species.

These equilibria are dynamic inasmuch as they shift with the changing conditions and according to the struggle for survival between species. The increase in the concentration of Solanaceae during intensive potato-growing periods brought about the overdevelopment of the Colorado beetle, resulting in its displacement far from its natural equilibrium position. If man had not intervened, the return to equilibrium would have occurred naturally following the fall in the number of Solanaceae plants. Cacti, which were artificially introduced by man to Australia, in the end drove him away from some fertile areas. This was stopped by introducing an insect that feeds on them. An 'insect–cacti' equilibrium was thus established and the number and distribution of these plants became limiting factors.

Among the examples of detrimental effects of man on nature, ocean intoxication has been chosen because it appears to be the most serious one in the long term. In natural environments, the subtlety of the interactions is such that minor modifications of one element in the system can eventually spread to the whole system. These considerations are valid for the marine environment if the observations are isolated geographically. But, on an oceanic scale, because of the immensity of the system, it is possible that equilibria can be re-established. The ocean mass behaves as an effective buffer against pollution. But, for the same reasons, the retention of the molecules which have been introduced is high because they disappear according to their own cycles. We shall briefly consider the problems brought about by the release of toxic waste products such as mercury or lead, the halogenous organic derivatives and oil hydrocarbons into the oceans.

The 'plant–insect' relationship

It is estimated that there are about 500,000 species of phytophagic insects. These include a large number which live on a single nutritive plant species or on a number of species belonging to the same family. Some of the plant metabolites can attract a given insect quite specifically, and the relationship is based on feeding habits or other specific needs. The same chemicals can be responsible for the attraction of the females which come to lay their eggs on a medium which is suitable for the development of the future larvae. The active scent is often a complicated mixture, and the chemicals which have been identified are usually only the main constituents. Coumarone of *Melilotus officinalis* attracts *Sitona cylindricollis* (Hans & Thorsteinson, 1961). Isothiocyanates and their glycoside precursors are responsible for the relationship between Cruciferae and Pieridae (as well as for *Listoderes obliquus*; Matsumoto, 1970). Honey-bees are attracted to clover pollen by the tri-unsaturated fatty acid **159** (Hopkins *et al.,* 1969). Propylmercaptan and propyldisulphate allow *Hylemya antiqua* to find its food (Matsumoto, 1970). The attractive power of the fly agaric *Amanita muscaria* is due to the 1,3-diester of glycerol with oleic acid (Muto & Sugawara, 1965, 1970). The insecticidal substances, which act after the initial attraction, have been described earlier. The meaning of this attraction, which is lethal for the flies, is unknown. The caterpillars of *Bombyx mori* recognize their food because of the presence of linalol and its acetate, citral and of terpinyl acetate. β-Sitosterol, morin, isoquercitrin, inositol and sucrose are phagostimulants (Hamamura, 1965). The Catalpa hawk moth, *Ceratomia catalpae* feeds on leaves containing catalposides and has a specific need of glucose (Nayar & Fraenkel, 1963). The coexistence of sugars is often necessary, as for instance in the case of the caterpillar of *Plutella maculipennis* whose feeding is induced by sinigrin **11**, but only in the presence of glucose (Thornsteinston, 1953). The cabbage aphid *Brevicoryne brassicae* can be made to feed on plants other than its natural choice if the stems are made to absorb sinigrin beforehand (Wensler, 1962). The beetle *Scolytus multistriatus* which lives on the American elm *Ulmus americana* is attracted by lupeyl cerotate (the ester of a fatty acid containing 26 carbon atoms and a pentacyclic triterpene (Dosckotch *et al.*, 1970)) and by (+)7-catechin β-D-xylopyranoside (Dosckotch *et al.*, 1973). The *Papilio ajax* caterpillars will only feed on umbellifers containing methylchavicol or carvone and may even accept a piece of paper soaked in carvone, devouring it as avidly as if it were an umbellifer leaf (Dethier, 1941). The males of the hymenopterans *Gorytes mystaceus* and *G. campestris* are attracted by the orchids of the *Ophrys* genus which produce a mixture of terpenes (including farnesol) which mimic the female sexual pheromone. The male attempts to mate with the flower and in this way the orchid is pollinated. As no other means for pollination exists in this plant, the survival of the species relies entirely on this curious relationship (Kullenberg, 1956).

According to Jermy (1966), the attraction of insects by plants to feed on them is complemented by the fact that plants which are not included in the normal diet of the insect contain substances which inhibit insect feeding (phagorepellents). It could be considered that many plants are protected by such repelling substances but that in the course of evolution certain insect species have become adapted to given plant families. Jermy's hypothesis is based on the fact that insects can in certain cases feed on completely synthetic diets whereas they cannot feed on all plants. Cucurbitacins, bitter and toxic substances found in Cucurbitaceae, attract the beetle *Diabotica undecimpunctata*. On the other hand, these tetracyclic triterpenes are repellents for bees and wasps (Chambliss & Jones, 1966). Hosozawa *et al.* (1964 a) isolated six diterpenes from the plant *Caryopteris divaricata* (clerodane skeleton, see clerodin **198**), which is an anti-feeding agent for the larvae of *Spodoptera littoralis*. These authors make a distinction between 'relative anti-feedants' which cause a delay but not an absolute inhibition of food intake, and 'absolute anti-feedants'. Bergaptene, obtained from *Beennighausenia albiflora* causes 100 per cent inhibition at a 500 p.p.m. concentration for 2 hours, but after 6 hours the insects begin to feed. Diterpenes of the clerodin type produce an irreversible refusal to feed. Another series of repelling terpenes has been isolated from Verbenaceae by Hosozawa *et al.*, (1974 b). Nic-2 **199**, a steroid which Bates and Morehead (1974) obtained from the solanacean *Nicandra physaloides*, was found to be an excellent phagorepellent. Nepetalactone **200**, isolated from the catnip *Nepeta cataria,* is a very active repellent (Eisner, 1964) related to the iridodial family and is used by ants for defensive purposes. The plant *Actinidia polygama* is protected from insects by a whole family of

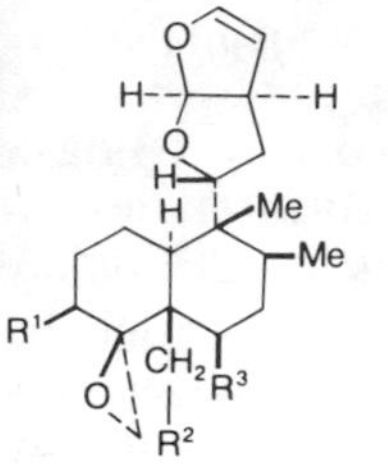

R¹ = H, R² = R³ = OAc

198

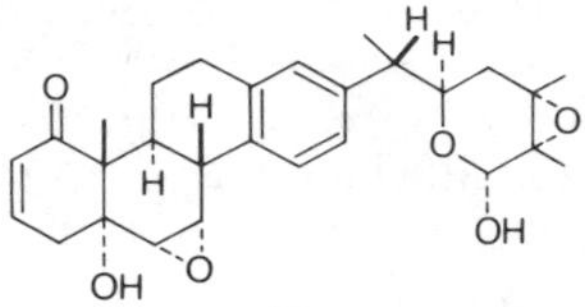

199

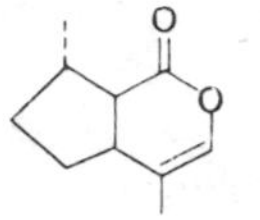

200

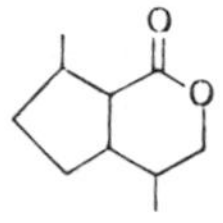

201

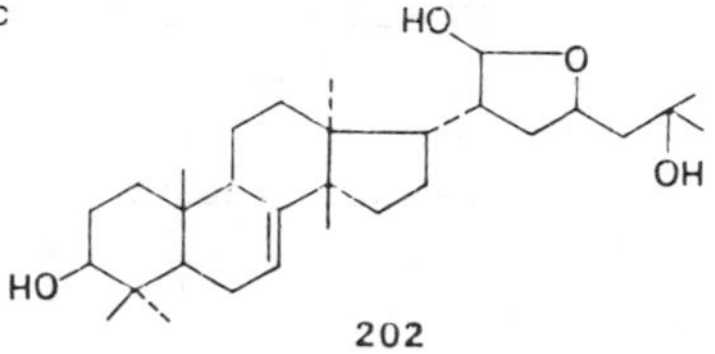

202

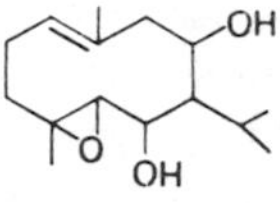

203

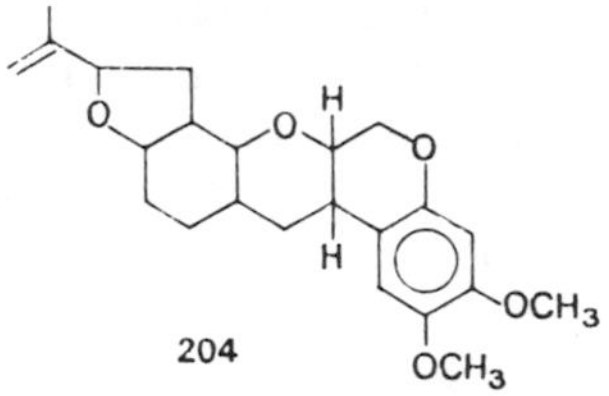

204

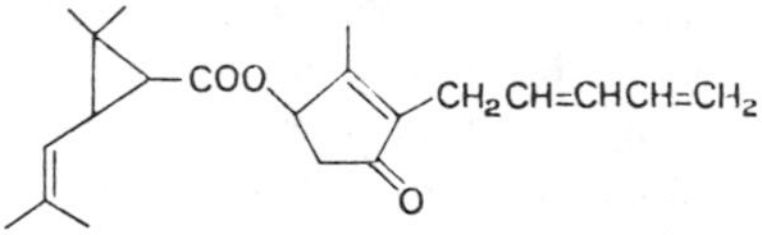

205

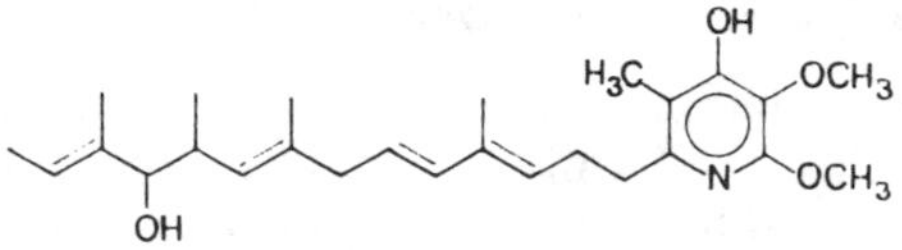

206

Fig. 26 Some molecules which play a role in plant–insect interactions.

iridoids which it is able to synthesize; among these is matabilactone **201** (Sakai *et al.,* 1959; Isoe *et al.,* 1968; Be Hyeon *et al.,* 1968).

On the other hand, *Chrysopa septempunctata*, which is not repelled by these substances, is instead attracted by them to *A. polygama* on which it feeds (see review by Herout, 1970). Nepetalactone appears to exert a strong attraction over cats which has yet to be explained. Some Meliaceae (*Melia azedarach* and *M. azadirachta*) are protected against locusts by meliantrol **202** (Lavie *et al.,* 1967). Wada *et al.* (1968 a, b) have shown that shiromodiol acetate **203** acts as an anti-feedant to *Prodenia litura*, a polyphagous insect, and to *Trimeresia miranda* which is oligophagous.

The ultimate means of plant defence is by the true insecticide, that is a chemical which, if ingested, causes death. In this context we may cite rotenone **204**, pyrethrin **205** and piericidin A **206** (Herout, 1970). We have already considered the effects on insects of substances mimicking juvenile hormones and of ecdysones.

The formation of galls by insects poses several interesting problems. Galls can appear on various parts of a plant's anatomy: new tissues show a complex structure and new differentiation appears. A single egg placed beneath a leaf is sufficient to trigger this neo-formation of tissues. Members of the following families can cause galls: Cynipidae (Hymenoptera), Cecidomyidae (Diptera), Tenthredinidae (Hymenoptera) and Psyllidae (Hemiptera). In the case of the formation of galls on willow leaves by *Pontania* (Hymenoptera), the egg does not have to be inserted into the plant tissues. The introduction of the ovipositor can be sufficient to produce a response. It is likely that a secretion is released during the egg-laying operation (Howanitz, 1959). Gall induction is a highly specific process, as any one *Pontania* species is able to produce galls in only one willow variety. It is likely that nucleic acid transmission by the insect is involved.

Carnivorous plants represent a peculiar example of plant–insect interaction. About 450 species of such plants are known. They are found in all climates and the design of the trap is extremely diverse. Prey is attracted by odoriferous substances or by nectars, and is either digested by enzymes or by bacteria. In Nepenthaceae, the proteolytic enzymes are secreted by specialized glands (up to 6000 per cm^2) and insect contact is the trigger for enzyme secretion.

The disturbance of oceanic equilibria by man

As we have already emphasized, the foundation of chemical ecology were firmly established by Lavoisier. The movement of molecules between the living world and the mineral world occurs through the phenomena of combustion and putrefaction, which are the main processes of restitution. The existence of cycles, those of carbon, nitrogen, sulphur, phosphorus, etc., relies entirely on the availability of entries, exits and exchanges of atoms and molecules between the mineral and animal worlds. In this context, the oceanic mass plays a determining role. The carbon cycle is based on the automatic regulation of atmospheric carbon dioxide. The system involving calcium carbonate, calcium bicarbonate and carbon dioxide is the buffer which is responsible for the maintenance of carbon dioxide levels. Carbon dioxide is removed from the atmosphere by photosynthesis and is restored through respiration, but even in this case, the ocean environment is the determining factor: photosynthesis is about eight times more intense in the seas than on the land (Florkin, 1956). This very high activity is no doubt due to the phytoplankton. The point of exit from the carbon cycle is the elimination of carbonates in the form of sediments. In the course of past centuries, significant quantities have been removed as fossil carbon. Since the beginning of life on earth, the total amount of carbon dioxide which has disappeared in the form of sediments is believed to correspond to about twelve times the total amount of carbon dioxide in the atmosphere at the present time (a considerable compensatory effect is due to volcanic eruptions). The role played by marine phytoplankton appears, therefore, to be a fundamental step, as well as the most sensitive one in the system. Human development has ceaselessly driven back vast areas of land vegetation, but it is now thought that ocean pollution is likely to alter the established equilibria in a much more profound way. Many examples of adaptation of the living world to its environment by specialization have been mentioned. In order to adapt, man made use of his intelligence and altered his environment. Consequently, he had to destroy the natural system which preceded him. Man's lack of foresight will induce very important modifications.

The exaggerated accumulation of nitrated waste products by rivers or town sewers can lead to progressive eutrophication. The first sign of eutrophication is the liberation of ammonia and the formation of nitrates. The resulting plant proliferation leads to oxygen impoverishment. This destruction of the environment is most clearly seen in some lakes and in coastal areas of land-locked seas.

A characteristic feature of phosphorus movement is its constant disappearance from continents through draining waters and its enrichment of the ocean depths where it accumulates. It is only reappearing very slowly, either by means of the primary production or through sea-birds. The phosphorus cycle is the critical point for biosphere metabolism.

Of all the detrimental influences which man exerts over his environment, the most damaging ones in the long term are those which affect the oceans. It is true that because of the mass of the oceans, the equilibria are not easy to shift, but by the same token, they are difficult to re-establish. Life, which probably comes from water and which relies on water, must respect water. The problem of the quality and quantity of water sources in the world is one of the most formidable which man will have to tackle in the next century. If Frenchmen were to use as much water as North Americans, half of the French rivers would have to be found in the taps. It is a well-known fact that the lack of water is responsible for the underdevelopment of a large part of the world, but if water consumption were normal – that is on the basis of water consumption in the USA – world population could not exceed 10 thousand million (Colas, 1973). In his treatise on the plague (Book XXII), Ambroise Pare wrote: 'The best water is rainwater; that of rivers is just as good, if taken below the surface, in the stream'. Times have changed and river waters continuously drag toxic loads towards the sea, and the accumulation of these is a mortgage over our future. If the waste water of the Paris region alone were correctly dealt with, a trainload of mud would have to be removed every day (Colas, 1973).

Pollution of the seas is not always obvious. The North Sea, for instance, in spite of being particularly affected by pollution, remained one of the most productive fishing grounds in the world (Cole, 1971).

The drop in production observed in certain areas is due more to the extensive use of modern fishing techniques than to the effects of pollution, at least in areas where tides are important. Pollution becomes much more apparent in estuaries and bays, which are the great nursery grounds for many species (Cole, 1971). Specialists who have studied the fauna and flora of the sea floor for several decades all agree that the intensity of life is being constantly diminished (Cousteau, 1970). Ocean intoxication is the result of many additive effects. In the paragraphs to follow, we shall briefly consider pollution caused by particular metals, by halogenous organic derivatives and by hydrocarbons (Ramade, 1974).

The food chain starts with phytoplankton, which can be described as a marine 'meadow'. As far as consumers are concerned, we first come across zooplankton, which may be herbivorous, carnivorous or omnivorous. Consumers of phytoplankton and zooplankton include a wide range of animals, marine invertebrates and fish for instance. The carnivorous chain will eventually lead to mammals including man. It is perhaps more realistic to view this system in terms of a branching network rather than a chain. At each level, toxic substances are being concentrated; and animals higher up in the chain accumulate substances found in those animals lower down. The concentrations are particularly high in the liver and in the lipid reserves.

Metals such as mercury, lead, cadmium, copper, zinc, chromium, etc., are among the most dangerous ones. The accumulation of mercury due purely to natural causes, namely the ceaseless drainage of the continents by rain, is already cause for concern. To this, the waste products originating in the combustion of fossil carbon resources, oils and coals, have to be added. Mercury is eliminated in the form of methylmercury and is very stable in a natural environment and will therefore be found in marine organisms (Fig. 27). The toxicity becomes apparent in the form of nervous disorders, even at very low concentrations. The accidents occurring in Japan (Minamata) between 1953 and 1960 are well known. During this period those affected were either killed outright or crippled by the consumption of crustacea or fish contaminated by methylmercury which had been released into the sea by a polyvinyl chloride factory. In man, methylmercury disappears slowly, with a half-life of 70 days. It can penetrate the placental wall and therefore affect the foetus. Normal mercury levels in fish are of the order of 0.01 to 0.1 mg/kg and can be higher in the vicinity of volcanoes (Eshleman *et al.*, 1971). Some fish species such as tuna and swordfish can accumulate naturally occurring mercury to harmful levels (the presence of high levels in these fish does not, therefore, indicate the presence of pollution; Grimstone, 1972). Tuna fish caught in the vicinity of coasts tend to contain more mercury; in fish caught in the Irish Sea the levels can be as high as 0.55 mg/kg (Grimstone, 1972). In certain seas (Japan, North America, Sweden) the mercury levels are close to 1 p.p.m. (the safe level is 0.5 p.p.m., with possible

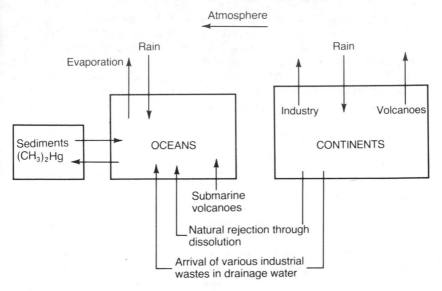

Fig. 27 Schematic representation of the mercury cycle.
Observed mercury concentrations in fresh water (according to Ramade, 1974):
Water 0.1 p.p.b. Aquatic insect larvae 0.1–1p.p.m. Phytoplanton 10–100
p.p.b. Microphagic fish 0.5–1 p.p.m. Zooplankton 100–500 p.p.b. Hunter
fish (pike) 4 p.p.m.

effects at 6 p.p.m.). Mercury originates from a variety of industrial
processes, including paper mills where organo-mercurials are used to
prevent bacterial growth in the paper pulp, chlorine and alkali
factories where electrolytic processes are used, and plastics factories
where mercury is used in catalysts. The transformation of mineral
mercury into methylmercury occurs essentially in sediments. The
sediments of the St Clair river in Canada, for instance, contain up to
1700 mg of mercury per kg (Grimstone, 1972).

Lead, which is found in the oceans, comes from land drainage, but to
a much larger extent from the atmosphere because of the use of the
internal combustion engine. The use of lead tetraethyl as an anti-knock
agent introduces 2×10^5 tonnes of lead every year. Surface waters and
sediments are the worst affected. In rainwater, concentrations as high
as 40 μg/l have been found, and in fog, as high as 300 μg/l of con-
densed water. In Great Britain in 1970, the amount of lead released
from the combustion of coal was estimated to be 120 tonnes (Bryce-
Smith, 1971 a, 1971 b, 1972; Mills, 1971). Metabolic disorders caused
by lead may arise at concentrations as low as 0.2 p.p.m. with the
toxic threshold being 0.3–0.5 p.p.m. In the case of lead tetraethyl,
the threshold for the inhibition of cell division in animal cells occurs
at a concentration which is lower than that for colchicine.

The presence of copper, zinc, cadmium and chromium introduced by rivers into the seas represents a risk for man because the invertebrates that he consumes (oysters, mussels, scallops, and so on) easily accumulate these metals. This ability of marine animals, more particularly filtering animals, to concentrate elements also holds for the concentration of radioactive elements, whether they arise from nuclear explosions or from industrial waste. Radioactive wastes, which amounted to 10,000 tonnes in 1958 and 100,000 tonnes in 1965, will reach 10 million tonnes per year by the year 2000 (Colas, 1973). Radioactive waste disposal is an extremely difficult problem to solve and the release of this type of waste into the oceans is not without considerable risk.

Organic chlorine derivatives, whether pesticides or various industrial by-products, are highly toxic and persistent. They have been found in the livers of animals which are at the upper end of the food chain, such as fish, penguins and seals. DDT has been extremely useful in the fight against malaria but it is believed that about 50 per cent of all the DDT ever produced is now in the oceans (Cole, 1971). Using the figures for world production, *Actualité Chimique* (1974) concludes that sea dilution results in a final concentration of 3 μg per m^3 and that it is inconceivable that any organisms can concentrate DDT starting from such low values. The phenomenon must, however, be assessed after taking local concentrations into account, for instance those in coastal waters and in surface waters (some deep waters are not renewed for long periods, centuries in some cases). The build-up in the concentrations of toxic chemicals in phytoplankton has indirectly resulted in the pollution of Antarctic penguins (George & Frear, 1966), which are at the end of the food chain. The increase in the concentration of chlorinated hydrocarbons at the sea surface is a well-documented phenomenon known as slick formation. Such areas sometimes contain up to 10,000 times more chlorinated hydrocarbons that neighbouring waters (US Bureau of Commercial Fisheries, 1970). These chemicals can enter fish directly through gill contact. The toxicity will vary according to temperature, water pH and other factors (Cole, 1971). Fish may become resistant to chlorinated products and can then accumulate even greater quantities. The study of polychlorodiphenyl slicks in the Sargasso Sea revealed that superficial layers (150 μm) contain 11.2 μg/l, whereas at a depth of 30 cm the concentration is only 3.6 μg/l (Bidleman & Olney, 1974). It is likely that one of the most damaging effects of chlorinated hydrocarbons is the inhibition of photosynthesis in phytoplankton. Such an effect is apparent at concentrations in the order of 10 parts per thousand million (Wurster, 1968). For a review of pesticide toxicity see Chasseaud (1970).

Estimates of the amount of oil lost at sea since the beginning of its transport by ship amount to about 5 million tonnes (Zo Bell, 1964; Pilpel, 1968). Oil pollution of the seas is not only caused through the actual transport of oil (leaks caused during offshore drilling operations are more common, for example) but the results tend to be more spectacular. The *Torrey Canyon* disaster will spring to most readers' minds. On 19 March 1967, this ship ran aground on the Seven Stones Reef, off the South-West of England, and more than 100,000 tonnes of crude oil were spilled. Yet even this terrible accident was dwarfed by that involving the tanker *Amoco Cadiz* off the Brittany coast in 1978. It is a mystery that oceans are not more polluted than they are!

Natural elimination processes come to bear upon spilt oil, including evaporation, dispersion by tides and waves, and destruction by oxidation or by bacterial consumption. Dispersion unfortunately brings oils to the coastlines or causes deposition over regions of natural sedimentation. Evaporation is an important factor where spills containing compounds with a boiling point below 150 °C are concerned. The lighter fractions soon evaporate and this explains why older slicks are not inflammable. The first effect of dispersion is the spreading of the slick into a thin layer, producing emulsions which are then absorbed by the ocean mass. The time period over which oil–water particle formation and its sedimentation occurs can be very long indeed; according to Pilpel (1968) it can last for several months at a depth of 4000 m. The oxidation of oil in air produces waxes and polymers (Nixon, 1972). The effect of ultraviolet light favours the formation of hydroperoxides which will either break down into smaller molecules or will polymerize. The breakdown of oil in the seas is thought to be accelerated by the activity of micro-organisms (Dzuban, 1958). Entry of the breakdown products of oil into the food chain begins at the point where protozoa feed on bacteria. All these various breakdown routes result in the sinking of surface oil. The end result is the appearance of small organic molecules, alcohol, acids, etc., which dissolve and enter the cycles of soluble organic matter. The use of industrial detergents slows down these natural processes by inhibiting bacterial proliferation (George, 1961). Anaerobic destruction occurs in the case of submerged oils and is a slower process. Hydrocarbons occur naturally in solution in sea water (Barbier, *et al.*, 1973) and are easily extracted from the medium by marine invertebrates, whether they are filter-feeders or not (Fevrier *et al.*, 1975).

A wide variety of synthetic organic compounds including phenols and detergents, are continually being released into drains and find their way to the sea. To these we must add the purposeful (but apparently rare) dumping of substances such as, for example, several thousand tonnes of a paralysing nerve gas (Sarin gas) in watertight containers by the US Army in the Gulf of Mexico in 1970 (Tam, 1970). Such dumping is based on the idea that containers remain forever waterproof.

The study of the distribution of plastic debris and particles on the surface of the Northern Atlantic, illustrates well the magnitude of oceanic pollution (Colton *et al.*, 1974). The indestructibility of plastics by the natural environment is leading to an accumulation to which there seems no limit. The diffusion of light constituents in water such as vinyl chloride or phthalates is widespread. Butyl and octyl phthalates, which are used as plasticizing agents, can be found in sea waters of various origins and at most depths (Copin & Barbier, 1971; Saliot, unpublished results). Fish may contain up to 3.2 mg/kg of these compounds (Mayer *et al.,* 1972). Pollution by these plasticizing agents is, in fact, a universal phenomenon; they have been found in plants, milk, fats, various foods and even in human blood (Hunnemam & Christ, 1974).

Future outlook

The imbalances created by *Homo sapiens* are not linked to his appearance on the earth but to the development of his social life. In small, isolated groups, primitive man had to endure the harsh conditions of his environment and could hardly be said to be involved in its modification. Natural equilibria were responsible for the regulation of his family size. This situation began to change when hunting in groups started, when land was cleared for cultivation and when fire was used for destructive purposes, whether for the conquest of new land, for its fertilization by ashes or for hunting. In this respect, the study of certain African tribes or Australian Aborigines is of great interest. We can consider man to have been in a state of equilibrium with his environment principally because of a variety of ecological factors. Social life is found in many species, particularly among insects, but these societies stopped evolving a long time ago. Fruit-picking, agriculture, breeding, war and sometimes slavery are encountered forms of social behaviour, but in a form which does not endanger the natural milieu. The honey-bee society can be compared to a machine, where the functions are controlled by pheromones which induce automatic reactions. The individual cannot live out of the group and if some form of social evolution does still exist, it must occur at such a slow pace that equilibria are given many opportunities to adapt. In the case of man, the development of individual awareness and the progressive discovery of techniques increased his power over his environment. The awareness and the respect of nature's values did not keep up with the development of intelligence. A confrontation rapidly developed and has become more marked in the last centuries. For instance the phenomenon of war still exists in our era of highly developed technology. The elaboration of new techniques and their rapid development can be regarded as a positive contribution to war itself. The exploitation of natural resources and the beginning of the industrial era were immediately followed by the exhaustion of these resources and an exaggerated accumulation of waste materials. Among the misfortunes of human societies, the effects of a well-known law appear, according to which any evolving system tends to give birth to forces which will resist its own evolution. For a long time, nature had been considered as an enemy and its destructive exploitation seemed to be a blessing rather than a misdeed.

Actually, any progress is ambiguous and the decisions will entirely depend on the mental level of the users. A pebble found in the road can as easily be a weapon as a tool; long before the discovery of iron or of the atom everything could be used for better as well as for worse. Considerable areas of cultivated land and of forest in Cambodia, Laos and Vietnam were sprayed with herbicides and defoliant chemicals, and this was just one aspect of chemical ecology; with chemical warfare, the discovery is used for the deliberate destruction of nature and mankind.

The identification and isolation of antibiotics is one of the greatest steps forward accomplished by the human species. The abolishment of the microbial mortgage is the onset of the infernal machine of overpopulation. Besides, the inconsiderate use of antibiotics has led to resistant strains, namely to new adaptations which cancel their effects. Penicillin, the wonder weapon against syphilis, has now become powerless in many cases. The increase in the average lifespan, which is also linked to better hygiene and social organization, is responsible for the appearance of cardiovascular diseases and cancers. Nature's riches are essentially still to be discovered but energy reserves such as coal and oil are on their way to exhaustion. The atom will take over, with all the risks involved. Since the timing has already been established, the accumulation of atomic waste is, to a certain extent, unavoidable. The ocean, for which rational exploitation plans still have to be formulated, is a fantastic source of food but its intoxication by man is well under way.

The progress of which man is capable always appears to be ambiguous. The ambiguity lies in the blunders, the lack of foresight, the lack of morality and also in the automatic responses.

Knowledge of the chemical interactions in nature has not led, up to now, to the possibility of accurate forecasting and this prospective aspect ought to be the subject of study in the very near future, because no effective control over the interactions of man and his environment is possible without previous knowledge of them.

The superabundance of the human species is in need of some form of control. India is a good example of the problems involved in population expansion. Mortality rates are very high there; it is said that about 80 per cent of the population is hungry. On the other hand, the yearly birth-rate is 4.3 per cent whereas the mortality rate is 1.8 per cent. By the year 2000, the population of India will reach a thousand million. In spite of the precarious life conditions, the average life expectancy of Indians has increased from 25 to 53 years. At present, the population is made up of 47 per cent children, 49 per cent adults and 4 per cent of old people. It is likely that by 2000, the young will

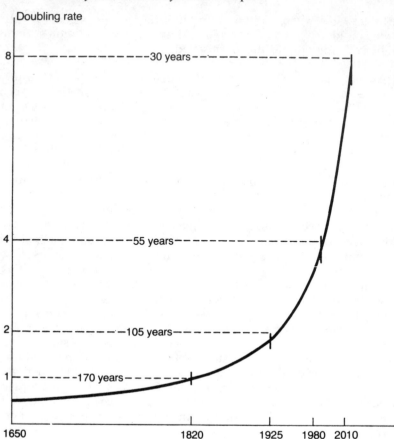

Fig. 28   World population expansion between 1950 and 2010 according to a study by the MIT.

represent 60 per cent of the total population (Bonnefous, 1968).

A study by the Massachusetts Institute of Technology illustrates the exponential character of world population growth (Fig. 28). It is hoped that rapid policies taken at national and international levels will prove these frightening predictions to be wrong, but it is difficult to imagine how the explosive character of this curve can be suppressed in much less than a century. The problem of hunger is linked to these pessimistic predictions. It is obvious that a large part of the world population will be subjected to the social modulations which are created by the dilemma: to die or to find food by any means. Cultivated areas cannot be increased at an adequate enough rate, yet the soil degradation brought about by intensive farming methods is quite apparent. To this picture, a number of other factors ought to be added, such as the drought in Sahel countries and in Ethiopia. It is not certain that the climatic changes observed since 1968 are purely fortuitous; it might be that they are the natural continuation of the phenomena responsible for the formation of the Sahara desert and just happen to be momentarily noticeable. The global picture which faces the ecologist is a rather demoralizing one. The compulsory use of birth control pills in overpopulated societies would be a new aspect of chemical ecology which would not be without analogies with what is known about the societies of the social insects.

The most useful results of chemical ecology are to have become aware of its own identity, to lead to an analysis of the situation, to have sounded the alarm and to have developed possible solutions. In these respects, chemical ecology is linked to ecology itself. As Bonnefous (1968) has remarked, solutions to grave human problems require preliminary action on the mental structures. Therefore, in order to become efficient, ecology must be the instigator of a new philosophy, but the effects are much too new and patience is required. In a survey of all the major world problems of his time Bertrand Russell (1962) considered, rather pessimistically, the future of humanity, but in his mind the real dangers were war and 'enrolment'. The latter term is used to express the idea that the coordination of scientific discoveries with bureaucratic development endangers mankind. According to Russell, a new organization of the world could be reached in which there is little room for freedom of thought. It must be realized that pollution and overpopulation, which already existed in 1962 but which were not then considered to be major problems, were absent from the preoccupations of Bertrand Russell. Awareness of them is therefore a recent development. It is even clear that the struggle against these problems could indeed be best fought by a coordination of science and administration.

Epilogue

'La psychologie de l'homme fait obstacle á la philosophie de l'homme.'
(Bachelard)

On the difficulty of concluding

The purpose of an epilogue is to collect together the partial conclusions developed during the book, to reconsider these in a more general framework and to discuss them. Faced with the difficulty of doing this, with the fear for the long-term future of humanity, it is hoped to end this book on an optimistic note.

Chemical ecology, the science of natural chemical interactions, is a broad attempt at understanding the living world. It is concerned with the effect of man on the biosphere and in this context the problems brought about by population are immediately relevant. Truly, the need to understand and to protect and the fear of destruction are two complementary notions which represent the same projection of dynamic thought. The attractiveness of a blue sky and of life is the mirror image of pollution and death. All human energy ought to be geared towards the quest for better ways to improve life and to fight death. However, death is a primary condition for life. It would therefore seem that to develop life is akin to increase in death potentials, including various forms of pollution. In this instance, the contradiction is not thought to be fundamental, but rather to be the mirror image of a progress vector. It might be felt that a certain degree of fatalism comes from considerations of this nature.

Chemical ecology aspires to its inclusion in man's philosophy which would help him to recover his correct place in his natural environment. But man's psychology tends to obstruct this. Among ecologists, the general feeling is not optimistic. Recalling a phrase by Ramade (1974), it is pressing to radically change the relationship between man and the biosphere if 'our species is to avoid the fate of dinosaurs'. In certain circles, ecologists are thought to be exaggerating, while others blame the scientists, witchcraft apprentices, who enabled the development of the industrial era without being able to foresee its consequences.

The destruction of the environment started with the first clearing of ground for cultivation and with the discovery of fire. It is linked to the first utterances of civilization. However, the industrial era caused a fantastic acceleration of these processes. To this picture, we must add the galloping nature of demography; the yearly increase in the human population was estimated to be 0.3 per cent in 1650 and had reached 2.1 per cent by 1970. It is believed that the world population will double every 33 years. Without mentioning the explosive character of this increase itself, it is obvious that pollution levels are linked to the demography, but the relationship is not linear. Indeed, pollution levels are lower than expected, on a planetary scale, because of the backwardness of underdeveloped countries, but this backwardness will be progressively buffered in the coming century. Could drastic measures, taken at national levels to begin with, slow down the infernal machine? Or is it to be believed that such measures will just result in a slowing down of the phenomenon without actually avoiding its final consequences?

Plato attempted to prove to us that knowledge and ethics were but one and the same thing. It is believed that chemical ecology and ecology itself lead to ethics, or rather are part of ethics and of a philosophy of man. We are led to predict the absolute necessity of changing rapidly man's psychology.

To improve our knowledge of nature and to develop an anti-pollution technology are the two complementary facets of such a policy. The study of the fate of organic molecules released in the environment will allow the determination of their life-cycles and an assessment of the conditions for their elimination. The switch from the use of insecticides as a means to eradicate pests to the use of more specific methods such as biological means, hormones or pheromones, should lead to pest control without pollution. DDT was extremely useful and saved millions of lives; thanks to its use, malaria has been eradicated in many areas. To improve on DDT is both a wish and an ambition of chemical ecology.

There are no miraculous cures for the misfortunes which are inflicted upon us and for those which lie in wait for us, but the will to rationalize man's actions is indeed a hope because it can lead to new planetary ethics.

Contemporary philosophers often proclaim that philosophy never has, and never will have, any positive results because as soon as it appears, it becomes the object of science. This is evidently a play on words, but nevertheless it remains that the essential aims are firstly to define what is being talked about and secondly to express the problems in clear terms. The real vocation of philosophy is therefore in the domain of language and expression. It then appears that some of the problems are imaginary and therefore do not deserve any further attention. In any case, they ought to be kept apart in case new elements of evidence are brought to light. A problem which is expressed in clear terms becomes available to all and can reach the collective conscience. It can bring about its own solution. Starting from words, an evolution motive has been created. The awareness, at the population scale, of the problems is absolutely indispensable for the achievement of solutions. Pollution problems are at the moment on the point of achieving this universal awareness and this is the hopeful note which had to be brought up. In this sense, ecologists achieved a reversal of the roles: the ecological science, which came out of a general philosophy of nature, is about to lead to a new philosophy of man.

References

Actualité chimique. (1974) Point de vue: les pesticides et l'environnement. Soc. Chim. Fr. n° **4**, 3–6.

AKRE, R. D. & RETTENMEYER, C. W. (1968) Trail following by guests of army ants. *J. Kansas, Entomol. Soc.,* **260**, 2334–5.

ALCAIDE, A. (1971) Sur les intermédiaires de la biosynthèse des phytostérols. *Thèse de Doctorat, Université d'Orsay,* 90 p.

ALCAIDE, A., DEVYS, M. & BARBIER, M. (1968) Remarques sur les stérols des algues rouges. *Phytochem.,* **7**, 329–30.

ALCAIDE, A., DEVYS, M. & BARBIER, M. (1969) Incorporation du desmostérol dans les stérols du tabac *Nicotiana tabacum. FEBS letters,* **3**, 257–9.

ALCAIDE, A., DEVYS, M., BOTTIN, J., FETIZON, M., BARBIER, M. & LEDERER, E. (1968) Sur la biosynthèse des stérols du tabac *Nicotiana tabacum* à partir du méthylène-24 cholestérol et du méthylène-24 dihydrolanostérol. *Phytochem.,* **7**, 1773–7.

ALDRIDGE, D. C., BURROWS, B. F. & TURNER, W. B. (1972) The structures of the fungal metabolites cytochalasins E and F. *Chem. Comm.,* 148–9.

ALDRIDGE, D. C. & TURNER, W. B. (1969) Structures of cytochalasins C and D *J. Chem. Soc.* (C), 923–8.

ALEXANDER, P. & BARTON, D. H. R. (1943) The excretion of ethylquinone by the flour beetle. *Biochem. J.,* **37**, 463.

ALLAIS, J. P. (1974) Les stérols chez les Insectes-Structures et origine des stérols du criquet *Locusta migratoria. Thèse de Doctorat, Université d'Orsay,* 135 p.

ALLAIS, J. P., ALCAIDE, A. & BARBIER, M. (1973) Fucosterol-24, 28 epoxide and 28-oxo β-sitosterol as possible intermediates in the conversion of β-sitosterol into cholesterol by the locust *Locusta migratoria. Experientia,* **29**, 944–5.

ALLAIS, J. P. & BARBIER, M. (1971) Sur les intermédiaires de la désalkylation en 24 du β-sitostérol par le criquet *Locusta migratoria. Experientia,* **27**, 506.

ALLAIS, J. P., PAIN, J. & BARBIER, M. (1971) La désalkylation en 24 du β-sitostérol par l'abeille *Apis mellifica. C. R. Acad. Sc.,* **272** (D), 877–9.

AMOURIQ, L. (1965) Origine de la substance dynamogène émise par *Lebistes reticulatus* femelle (Poisson Poeciliidae, Cyprinodontiformes). *Comptes-rendus Acad. Sci.,* **260**, 2334–5.

ANASTASI, A., ERSPAMER, V. & BUCCI, M. (1971) Isolation and structure of bombesin and alytesin, two analogous active peptides from the skin of the European amphibians *Bombina* and *Alytes. Experientia,* **27**, 166–7.

ANASTASI, A., ERSPAMER, V. & CEI, J. M. (1964) Isolation and amino-acid sequence of physalaemin, the main active polypeptide of the skin of *Physalaemus fuscumaculatus. Arch. Biochem. Biophys.,* **108**, 341–8.

ANASTASI, A., ERSPAMER, V. & ENDEAN, R. (1968) Isolation and amino-acid sequence of caerulein, the active decapeptide of the skin of *Hyla caerulea. Arch. Biochem. Biophys.,* **125**, 57–68.

ANDERSSON, C. O., BERGSTROM, G., KULLENBERG, B. & STALLBERG-STENHAGEN, S. (1967) Identification of macrocycle lactones as odiferous components of the scent of the solitary bees *Halictus calceatus* and *H. albipes. Arkiv f. Kemi,* **26**, 191–8.

ANESHANSLEY, D. J., EISNER, T., WIDOM, J. M. & WIDOM, B. (1969) Biochemistry at 100 °C: explosive secretory discharge of bombardier beetles (*Brachynus*). *Science,* **165**, 61–3.

APLIN, R. T., BENN, M. H. & ROTHSCHILD, M. (1968) Poisonous alkaloids in the body tissues of the cinnabar moth *Callimorpha jacobea*. *Nature,* 219, 747–8.

VON ARDENNE, M., OSSKE, G., SCHREIBER, K., STEINFELDER, K. & TÜMMLER, R. (1965) Sterine und Triterpenoide VIII. Uber die Sterine von *Solanum tuberosum. Die Kulturpflanze,* 13, 101–13.

Ibid. IX Uber die Sterine und Triterpenoide von *Solanum* und *Solanum polyadenium.* op. cit. 115–29.

ARPINO, P., VAN DORSSELAER, A., SERVIER, K. D. & OURISSON, G. (1972) Cires aériennes dans une forêt de pins. *C. R. Acad. Sc.,* 275 (D), 2837–40.

ARSENAULT, G. P., BIEMANN, K., BARKSDALE, A. W. & MCMORRIS, T. C. (1968) The structure of antheridiol, a sex hormone in *Achlya bisexualis. J. Amer. Chem. Soc.,* 90, 3635–6.

ASHLEY, C. C. (1971) Le calcium et l'activation du muscle squelettique. *Endeavour (Fr. edn),* 30, 18–25.

ASSELINEAU, J. & ZALTA, J. P. (1973) *Les antibiotiques; structure et exemples de mode d'action.* Hermann ed., Paris, 364 p.

AUBERT, M. (1971) Télémédiateurs chimiques et équilibres biologiques océaniques. I-Théorie générale. *Revue Intern. Oceanog. Médicale,* 21, 5–16.

AUBERT, M., PESANDO, D. & PINCEMIN, J. M. (1972) Télémédiateurs chimiques et équilibre biologique océanique. IV. Seuil d'activité de l'inhibition de la synthèse d'un antibiotique produit par une diatomée. *Rev. Intern. Oceanog. Med.,* 25, 17–22.

AUSTIN, D. J., BU'LOCK, J. D. & GOODAY, G. W. (1969) Trisporic acids: sexual hormones from *Mucor mucedo* and *Blakeslea trispora, Nature,* 223, 1178–9.

BALOZET, L. (1971) Scorpionism in the old world. In BÜCHERL, W. (ed.) vol. III, 349–71.

BANNER, A. H. (1974) The biological origin and transmission of ciguatoxin. In HUMM, H. J. & LANE, C. E. *Bioactive compounds from the sea.* M. Dekker Inc., New York, 15–36.

BARBIER, M. (1959) Séparation de p-benzoquinones par chromatographies sur plaques. *J. Chromatogr.,* 2, 649.

BARBIER, M. (1966) Le méthylène-24 cholestérol. *Ann. Abeilles,* 9, 243–9.

BARBIER, M. (1968) *Biochimie de l'Abeille;* in *Traité de l'Abeille.* Masson, Paris, vol. 1, 378–409.

BARBIER, M. (1972) *The chemistry of some amino-acid derived phytotoxins.* In WOOD R. K. S. *et al.,* 91–103.

BARBIER, M. & HÜGEL, M. F. (1961) Synthèses dans la série de l'acide céto-9 décène-2 *trans* oïque (Substance Royale). *Bull. Soc. Chim. Fr.,* 951–4.

BARBIER, M., JOLY, D., SALIOT, A. & TOURES, D. (1973) Hydrocarbons from sea water. *Deep-sea Research,* 20, 305–14.

BARBIER, M. & LEDERER, E. (1957) Sur les benzoquinones du venin de trois espèces de myriapodes. *Biochimia,* 22, 236–40.

BARBIER, M. & LEDERER, E. (1960) Structure chimique de la Substance Royale des reines d'Abeilles. *C. R. Acad. Sc.,* 250, 4467–9.

BARBIER, M., LEDERER, E., REICHSTEIN, T. & SCHINDLER, O. (1960) Auftrennung der sauren Anteile von Extrakten aus Bienen Königinnen. Isolierung des als Königinnen Substanz bezeichneten Pheromones. *Helv. Chim. Acta,* 43, 1682–9.

BARKSDALE, A. W. (1969) Sexual hormone of *Achlya* and other fungi. *Science,* 166, 831–7.

BARTELL, R. J. & LAURENCE, L. A. (1973) Reduction in responsiveness of males of *Epiphyas postvittana (Lepidoptera)* to sex pheromone, following previous brief pheromone exposure. *J. Insect Physiol.,* 19, 845–55.

BARTELS-KEITH, J. R. (1960) Alternaric acid. Part II. Oxidation. *J. Chem. Soc.,* 860–6. Part III. *Structure,* 1662–5.

BARTELS-KEITH, J. R. & TURNER, W. B. (1960) Alternaric acid. Part IV. *J. Chem. Soc.,* 3413–15.

BARTHEL, W. F., GREEN, N., KEISER, I. & STEINER, L. F. (1957) Anisylacetone, synthetic attractant for male melon fly. *Science,* 126, 654.

BASLOW, M. H. (1969) *Marine pharmacology.* William & Wilkins, Baltimore, 286 p.

BATES, R. B. & MOREHEAD, S. R. (1974) Structure of Nic-2, a major steroidal constituent of the insect repellent plant *Nicandra physaloides. J.C.S. Chem. Comm.,* 125–6.

BATISTE, B. C. (1972) Integrated control of codling moth on pears in California. A practical consideration where moth activity is under surveillance. *Environ. Entomol.,* **1,** 213–18.

BAUMANN, H. (1974) The isolation, partial characterization and bio-synthesis of the paragonial substances PS-1 and PS-2 of *Drosophila funebris. J. Insect Physiol.,* **20,** 2181–94.

BAUMANN, H., WILSON, K. J., CHEN, P. S. & HUMBEL, R. E. (1975) The amino-acid sequence of a peptide (PS-1) from *Drosophila funebris*: a paragonial peptide from males which reduces the receptivity of the female. *Eur. J. Biochem.,* **52,** 521–9.

BE HYEON, S., ISOE, S. & SAKAN, T. (1968) The structure of neo matatabiol, the potent attractant for *Chrysopa* from *Actinidia polygama. Tetrahedron. letters,* 5325–6.

BEAN, G. A. (1973) Phytosterols. *Advances Lipids Res.,* **2,** 193–217.

BECKER, G. (1964) Termite-anlockende Wirkung einiger bei Basidomyceten Angriff in Holz enstehender Verbindungen. *Holzforschung,* **18,** 168–72.

BEHAL, M. & PHISALIX, M. (1900) La quinone, principe actif du venin du *Julus terrestris. Bull. Mus. Hist. Nat.,* **6,** 388–90.

BENEDICT, R. G. (1972) Mushroom toxins other than *Amanita*. In KADIS, S. *et al.,* VIII, 281–320.

BENNETT, R. D. & HEFTMANN, E. (1965) Biosynthesis of *Holarrhena* alkaloids from pregnenolone and progesterone. *Phytochem.,* **4,** 873–9.

BERGSTRÖM, G. (1973) Chemistry of behaviour releasing olfactory signals in aculeate hymenoptera. *Abstracts of Gothenburg Dissertations in Science,* **34,** 36 p.

BERGSTRÖM, G., KULLENBERG, B., STÄLLBERG-STENHAGEN, S. & STENHAGEN, E. (1968) Identification of a 2, 3-dihydrofarnesol as the main component of the marking perfume of male bumble bee species *Bombus terrestris. Arkiv f. kemi,* **28,** 453–69.

BERGSTRÖM, G. & LÖFQVIST, J. (1970) Chemical basis for odour communication in four species of *Lasius* ants. *J. Insect Physiol.,* **16,** 2353–5.

BERGSTRÖM, G. & LÖFQVIST, J. (1971) *Camponotus ligniperda,* a model for the composite volatile secretions of Dufour's gland in formicine ants. In TAHORI, A. S. (ed.), Chemical releasers in Insects. *Proceed. 2nd Intern. IUPAC Congr. Tel-Aviv, 3,* Gordon & Breach, New York, 195–223.

BERGSTRÖM, G. & LÖFQVIST, J. (1972) Similarities between the Dufour's gland secretions of the ants *Camponotus ligniperda* and *C. herculeanus. Entomol. Scand., 3,* 225–38.

BERGSTRÖM, G. & LÖFQVIST, J. (1973) Chemical congruence of the complex odoriferous secretions from Dufour's gland in the species of ants of the genus *Formica. J. Insect Physiol.,* **19,** 877–907.

BERKOFF, C. E. (1971) Insect hormones and insect control. *Chem. Educ.,* **48,** 577–81.

BEROZA, M. (1970) *Chemicals controlling insect behaviour.* In BEROZA, M. (ed.), Academic Press, New York, 145–50.

BEROZA, M., ALEXANDER, B. H., STEINER, L. F., MITCHELL, W. C. & MUYASHITA, D. H. (1960) New synthetic lures for the male melon fly. *Science,* **131,** 1044–5.

BETHEL, R. S., FALCON, L. A., BATISTE, W. C., MOREHEAD, G. W. & DELFINO, E. P. (1972) Sex pheromone traps aid in determining need for codling moth control in apples and pears. *California Agric.,* **26,** 10–12,

BEVANS, C. W. L., BIRCH, A. J. & CASWELL, H. (1961) An insect repellent from black cocktail ants. *J. Chem. Soc.,* 488.

BIDLEMAN, T. F. & OLNEY, C. E. (1974) Chlorinated hydrocarbons from Sargasso Sea atmosphere and surface water. *Science,* **183,** 516–17.

BIERL, B. A., BEROZA, M. & COLLIER, C. W. (1970) Potent sex attractant of the gypsy moth: its isolation, identification and synthesis. *Science,* **170,** 87–9.

BIRCH, A. J., BROWN, W. V., CORRIE, J. E. T. & MOORE, B. P. (1972) Neocembrene A, a termite trail pheromone. *J.C.S. Perkin Trans.* **I,** 2653–8.

102   References

BIRCH, A. J., McDONALD, P. L. & POWELL, V. H. (1970) Reactions of cyclohexadienes. Part VIII. Stereoselective and non-stereoselective syntheses of ± juvabione. *J. Chem. Soc.* (C)., 1469–76.

BIRCH, M. C. (1974) Pheromones. *Frontiers of Biology*, **32**, North-Holland Research Monographs 595 p.

BISSET, G. W., FRAZER, J. F. D., ROTHSCHILD, M. & SCHACHTER, M. (1960) A pharmacologically active choline ester and other substances in the garden tiger moth *Arctia caja*. *Proc. Roy. Soc. London*, **B152**, 255–62.

BLINKS, L. R. (1951) *Physiology and biochemistry of algae*. In SMITH, G. M., *Manual of phycology;* Chronica Botanica Co., Waltham, New York, p. 66.

BLUM, M. S. (1969) Alarm pheromones. *Ann. Rev. Entomol.*, **14**, 57–80.

BLUM, M. S., (1974) *Pheromonal society* in Hymenoptera. In BIRCH, M. C., 222–49.

BLUM, M. S., CREWE, R. M., KERR, W. E., KEITH, L. H., GARRISON, A. W. & WALKER, M. M. (1970) Citral in stingless bees; isolation and functions in trail laying and robbing. *J. Insect Physiol.*, **16**, 1637–48.

BLUM, M. S., CREWE, R. M., SUDD, J. H. & GARRISON, A. W. (1969) 2-hexenal, isolation and function in *Crematogaster* species. *J. Ga. Entomol. Soc.*, **4**, 145–8.

BOCH, R. & SHEARER, D. A. (1962a) Identification of geraniol as the active component in the Nasanoff pheromone of the honeybee. *Nature*, **194**, 704–6.

BOCH, R., SHEARER, D. A. & STONE, B. C. (1962) Identification of iso-amylacetate as an active component in the sting pheromone of the honeybee. *Nature*, **195**, 1018–20.

BODANSKY, M. & PERLMAN, D. (1969) Peptide antibiotics. *Science*, **163**, 352–8.

BÖHNER, B. & TAMM, C. (1966) Die Konstitution von Poridin A. *Helv. Chim. Acta*, **49**, 2527–46.

BONNEFOUS, E. (1968) *Le monde est-il surpeuplé?* Hachette, Paris, 239 p.

BONNER, J. (1950) The role of toxic substances in the interaction of higher plants. *Botan. Rev.*, **16**, 51–65.

BOQUET, P. (1966) Venins de serpents. 2e partie. *Toxican*, **3**, 243–79.

BORDEN, J. H. (1974) Aggregation pheromones in the *Scolytidae*. In BIRCH, M C., 135–60.

BOUSQUET, J. F. & BARBIER, M. (1972) Sur l'activité phytotoxique de trois souches de *Phoma exigua* et la présence de cytochalasine B (phomine) dans leur milieu de culture. *Phytopathol. Zeit.*, **75**, 365–7.

BOUTRY, J. L. & BARBIER, M. (1974) La diatomée marine *Chaetoceros simplex calcitrans* et son environnement. I. Relation avec le milieu de culture; étude de la fraction insaponifiable, des stérols et des acides gras. *Marine chem.*, 217–27.

BOWERS, W. S., FALES, H. M., THOMSON, M. J. & UEBEL, E. C. (1966) Juvenile hormone: identification of an active compound from balsam fir. *Science*, **154**, 1020–1.

BOWERS, W. S., NAULT, L. R., WEBB, R. E. & DUTKY, S. R. (1972) Aphid alarm pheromone isolation, identification, synthesis. *Science*, **177**, 1121–2.

BOYLAN, D. B. & SCHEUER, P. J. (1967) Pahutoxin: a fish poison. *Science*, **155**, 52–6.

BRAND, J. M., BRACKE, J. W., MARKOVETZ, A. J. & WOOD, D. L. (1975) Production of verbenol by a bacterium isolated from bark beetles. *Nature*, **254**, 136–7.

BROCKMANN, H., von FALKENHAUSEN, E. H., NEEFF, R., DORLARS, A. & BUDDE, G. (1951) Die Konstitution des Hypericins. *Chem. Ber.*, **84**, 865–87.

BROOKSBANK, B. W. L., BROWN, R. & GUSTAVSON, J. A. (1974) The detection of 5 α androstene-16 ol-3 α in human male axillary sweat. *Experientia*, **30**, 864–5.

BROWN, W. L., EISNER, T. & WHITTAKER, R. H. (1970) Allomones and Kairomone: transpecific chemical messengers. *Biosciences*, 21–2.

BROWNLEE, R. G., SILVERSTEIN, R. N., MÜLLER-SCHWARZE, D. & SINGER, A. G. (1969) Isolation, identification and function of the chief component of the male tarsal scent in black-tailed deer. *Nature*, **221**, 284–5.

BRUCE, J. M. (1966) Smell as an exteroceptive factor. *J. Animal Sc.*, **25**, Suppl., 83–9.

BRYCE-SMITH, D. (1971 a) Lead pollution, a growing hazard to public health. *Chem. Brit.*, **7**, 54–6.

BRYCE-SMITH, D. (1971 b) Lead pollution from petrol. *Chem. Brit.*, **7**, 284–6.

BRYCE-SMITH, D. (1972) Behavioural effects of lead and other heavy metal pollutants.

Chem. Brit., **8**, 240–3.

BÜCHERL, W. (ed.) (1968–72) *Venomous animals and their venoms*, vols. 1–111. Academic Press, New York and London.

BÜCHERL, W. (1968) Introduction. In BÜCHERL, W. (ed.) vol. I, ix–xii.

BÜCHERL, W. (1971) Classification, biology and venom extraction of scorpions. In BÜCHERL, W. (ed.), vol. III, 317–47.

BÜCHERL, W. (1971) Venomous chilopods and centipedes. In BÜCHERL, W. (ed.), vol. III, 169–96.

BU'LOCK, J. D., JONES, B. E., QUARRIE, S. A. & WINSKILL, N. (1973) The biochemical basis of sexuality in *Mucorales. Naturwiss.*, **60**, 550–1.

BU'LOCK, J. D., JONES, B. E. & WINSKILL, N. (1974) Structures of the mating-type-specific prohormones of *Mucorales. J.C.S. Chem. Comm.*, 708–9.

BURTT, E. (1947) Exudate from millipedes with particular reference to its injurious effects. *Tropical diseases Bull.*, **44**, 7–12.

BUTENANDT, A. (1959 a) Wirkstoffe des Insektenreiches. *Naturwiss.*, **46**, 462–73.

BUTENANDT, A. (1959 b) Geschlechtsspezifische Lockstoffe der Schmetterlinge. *Jahrbuch 1959 der Max Planck Gesellschaft zur Forderung der Wissenschaften*, 23–33.

BUTENANDT, A., BECKMAN, R. & STAMM, D. (1961) Über den Sexuallockstoff des Seidenspinners II. Konstitution und Konfiguration des Bombykols. Hoppe-Seyler's *Z. Physiol. Chem.*, **324**, 84–7.

BUTENANDT, A., HECKER, H., HOPP, M. & KOCH, W. (1962) Die Synthese des Bombykols und der *cis-trans* Isomeren Hexadecadien-10, 12 ole-1. *Ann.* **658**, 39 64.

BUTENANDT, A. & KARLSON, P. (1954) Über die Isolierung eines Metamorphose-Hormons der Insekten in kristallisierter Form. *Z. Naturforsch.*, **9 b**, 389–91.

BUTENANDT, A. & TAM, N. D. Über einen geschlechtsspezifischen Duftstoff der Wasserwanze *Belostoma indica (Lethocerus indicus)*. Hoppe-Seyler's *Z. Physiol. Chem.*, **308**, 277–83.

BUTLER, C. G. (1967) Insect pheromones. *Biol. Rev.*, **42**, 42–87.

BUTLER, C. G. (1970) Chemical communications in insects: behavioural and ecological aspects. In JOHNSTON, J. W., MOULTON, D. G. & TURK, A. (eds) *Advances in chemoreception*, Appleton Century Crofts, New York, 35–70.

BUTLER, C. G. & CALAM, D. H. (1969) Pheromones of the honeybee. The secretion of the Nassanoff gland of the workers. *J. Insect Physiol.*, **15**, 237–44.

CAILLOIS, R. (1963) *Le mimétisme animal*. Hachette, Paris, 106 p.

CALABY, J. H. (1968) *The* Platypus *and its venomous characteristics*. In BÜCHERL, W. (ed.) vol. I, 15–29.

CALAM, D. H. & YOUDEOWEI, A. (1968) Identification and functions of secretions from the posterior scent gland of the 5th instar larvae of the bug *Dysdercus intermedius. J. Insect Physiol.*, **14**, 1147–58.

CALLOW, R. K. & JOHNSTON, R. C. (1960) The chemical constitution and synthesis of queen substance of honeybees. *Bee World*, **41**, 152–3.

CANONICA, L., FIECCHI, A., GALLI-KIENLE, M. & SCALA, A. (1966) The constitution of cochliobolin. *Tetrahedron letters*, 1211–18.

CARDANI, C., GHIRINGHELLI, D., MONDELLI, R. & QUILICO, A. (1965) The structure of pederin. *Tetrahedron*, **29**, 2537–45.

CARDÉ, R. T., ROELOFS, W. L. & DOANE, C. C. (1973) Natural inhibitor of the gypsy moth sex attractant. *Nature*, **241**, 474–5.

CARLSON, D. A., MAYER, M. S., SILHACEK, D. L., JAMES, J. D., BEROZA, M. & BIERL, B. A. (1971) Sex attractant pheromone of the house fly: isolation, identification and synthesis. *Science*, **174**, 76–8.

CARTER, F. L., DINUS, L. A. & SMYTHE, R. V. (1972) Effect of wood decayed by *Lenzites travea* on the fatty acid composition of the eastern subterranean termite *Reticulitermes flavipes. J. Insect Physiol.*, **18**, 1387–95.

CARTER, S. B. (1972) Les cytochalasines, outil de recherche en cytologie. *Endeavour (Fr. edn)*, **31**, 77–82.

CASINOVI, C. G. (1972) *Terpenoid phytotoxins*. In WOOD R. K. S. *et al.*, 105–25.

CASTELLANI, A. A. & PAVAN, M. (1966) Prime ricerche sulla biogenesa della

dendrolasina. *Bull. Soc. Ital. Biol. Sper.*, **42**, 221.
CAVILL, G. W. K., FORD, D. L. & LOCKSLEY, H. D. (1956) Iridodiol and iridolactone. *Chem. Ind.*, 465.
CAVILL, G. W. K. & HINTERBERGER, H. (1960) The chemistry of ants. Terpenoid constituents of some *Dolichoderus* and *Iridomyrmex*. *Austral. J. Chem.*, **13**, 514–19.
CAVILL, G. W. K. & HINTERBERGER, H. (1961) The chemistry of ants. Structure and reactions of dolichodial. *Austral. J. Chem.*, **14**, 143–9.
CAVILL, G. W. K. & HINTERBERGER, H. (1963) Dolichoderine ant extractives. *Proceed. 11th Intern. Congr. Entomol. Vienne*, **3**, 53–9.
CAVILL, G. W. K. & ROBERTSON, P. L. (1965) Ant venoms, attractants and repellents. *Science*, **149**, 1337–45.
CHAMBLISS, O. L. & JONES, C. M. (1966) Cucurbitacins: specific insect attractants in Cucurbitacae. *Science*, **153**, 1392–3.
CHANLEY, J. D., MEZETTI, T. & SOBOTKA, H. E. (1966) The holothurinogenins. *Tetrahedron*, **22**, 1857–84.
CHANLEY, J. D., PERLSTEIN, J., NIGRELLI, R. F. & SOBOTKA, H. (1960) Further studies on the structure of holothurin. *Ann. N. Y. Acad. Sc.*, **90**, 902–9.
CHANLEY, J. D. & ROSSI, C. (1969) The holothurinogenins II. Methoxylated neo-holothurinogenins. *Tetrahedron*, **25**, 1897–909. III. Neo-holothurinogenins by enzymatic hydrolysis of desulfated holothurin A., ibid., 1911–20.
CHASSEAUD, L. F. (1970) Foreign compound metabolism in mammals. *Chem. Soc. Lond.*, **1**, 75–87.
CHEYMOL, J., BOURILLET, F. & ROCH-ARVEILLER, M. (1970) Le terme toxine a-t-il encore une signification précise? *Toxicon*, **8**, 85–8.
CHOUSSY, M. & BARBIER, M. (1973) Pigments biliaires de Lépidoptères: identification de la phorcabiline I et de la sarpédobiline chez diverses espèces. *Biochem. systematics*, **1**, 199–201.
CIEGLER, A., DETROY, R. W. & LILLEHOJ, E. B. (1971) Patulin, penicillic acid and other carcinogenic lactones. In KADIS, S. *et al.*, vol. VI, 409–34.
CLARK, A. J. & BLOCH, K. (1959) Conversion of ergosterol to 22-dehydro-cholesterol in *Blatella germanica*. *J. Biol. Chem.*, **234**, 2589–94.
CLAYTON, R. B. (1960) The role of intestinal symbionts in the sterol metabolism of *Blatella germanica*. *J. Biol. Chem.*, **235**, 3421–5.
CLAYTON, R. B. (1964) The utilization of sterols by insects. *J. Lipids Res.*, **5**, 3–19.
CLAYTON, R. B. (1965) Biosynthesis of sterols, steroids and terpenoids. Part I. Biogenesis of cholesterol and the fundamental steps in terpenoid biosynthesis. *Quarterly Rev.*, **19**, 168–200. Part II. Phytosterols, terpenes and the physiologically active steroids, ibid., 201–30.
CLAYTON, R. B. & BLOCH, K. (1963) Sterol utilization in the beetle *Dermestes vulpinus*. *J. Biol. Chem.*, **238**, 586–91.
CLAYTON, R. B. & EDWARDS, A. M. (1961) The essential cholesterol requirement of the roach *Eurycotis floridana*. *B. B. Res. Comm.*, **6**, 281–4.
CLIVE, D. L. (1968) Chemistry of tetracyclines. *Quarterly Rev.*, **22**, 435–6.
COLAS, R. (1973) *La pollution des eaux*. Coll. Que-Sais-je. Presses Universitaires de France, Paris, 126 p.
COLE, H. A. (1971) Pollution of the seas. *Chem. Brit.*, **7**, 232–5.
COLLIGNON-THIENNOT, F., ALLAIS, J.-P. & BARBIER, M. (1973) Existence de deux voies de biosynthèse du cholestérol chez un mollusque, la Patelle. *Patella vulgata*. *Biochimie*, **55**, 575–82.
COLTON, J. B., KNAPP, F. D. & BURNS, B. R. (1974) Plastic particles in surface waters of the northern Atlantic. *Science*, **185**, 491–7.
COMFORT, A. (1971) Likelihood of human pheromones. *Nature*, **230**, 432–4.
COPE, O. B. (1971) Interaction between pesticides and wildlife. *Ann. Rev. Entomol.*, **16**, 325–64.
COPIN, G. & BARBIER, M. (1971) Substances organiques dissoutes dans l'eau de mer. Premiers résultats de leur fractionnement. *Cahiers océanographiques*, n° **5**, 455–64.
CORMIER, M. J., HORI, K. & ANDERSON, J. M. (1974) Bioluminescence in coelenterates.

Biochim. Biophys. Acta, **346**, 137–64.
CORMIER, M. J., WAMPLER, J. E. & HORI, K. (1973) Bioluminescence: chemical aspects. *Fortschr. Chem. Org. Naturstoffe*, **30**, 1–60.
COTT, H. B. (1938) Camouflage in nature and in war. *Royal Engin. J.*, 501–17.
COUSTEAU, C. (1970) Le problème des pollutions des mers. *Information Conseil de l'Europe*, 1–9.
CRONE, H. D. & KEEN, T. E. B. (1971) Further studies on the biochemistry of the toxins from the sea wasp *Chironex fleckeri*. *Toxicon*, **9**, 145–51.
DADD, R. H. (1960) The nutritional requirement of locusts. II. Utilization of sterols. *J. Insect Physiol.*, **5**, 161–8.
DAGLISH, C. (1950) The isolation and identification of a hydrojuglone glycoside occuring in the walnut. *Biochem. J.*, **47**, 452–7.
DAHM, K. H., RÖLLER, H. & TROST, B. M. (1968) The juvenile hormone IV. Stereo-chemistry of juvenile hormone and biological activity of some of its isomers and related compounds. *Life Sciences*, **7**, 129–37.
DAHM, K. H., BROST, B. M. & RÖLLER, H. (1967) The juvenile hormone V. Synthesis of the racemic juvenile hormone. *J. Amer. Chem. Soc.*, **89**, 5292–4.
DALY, J. W., TOKUYAMA, T., HABERMEHL, G., KARLE, I. L. & WITKOP, B. (1969) Froschgift. Isolierung und struktur von Pumiliotoxin C. *Ann. Chem.*, **729**, 198–202.
DALY, J. W. & WITKOP, B. (1971) Chemistry and pharmacology of frog venoms. In BÜCHERL, W. (ed.), vol. II, 497–519.
DELEUIL, G. (1950) Mise en évidence de substances toxiques pour les thérophytes dans les associations du rosmarino-ericion. *C. R. Acad. Sc.*, **230**, 1362–4.
DELEUIL, G. (1951 a) Origine des substances toxiques du sol des associations sans thérophytes du rosmarino-ericion. *C. R. Acad. Sci.*, **232**, 2038–9.
DELEUIL, G. (1951 b) Explication de la présence de certains thérophytes rencontrés parfois dans les associations rosmarino-ericion. *C. R. Acad., Sc.*, **232**, 2476–7.
DELEUIL, G. (1954) Action réciproque et interspécifique des substances toxiques radiculaires. *C. R. Acad. Sc.*, **238**, 2185–6.
DE LUCA, P., DE ROSA, M., MINALE, L. & SODANO, G. (1972) Marine sterols with a new pattern of side-chain alkylation from the sponge *Aplysina aerophoba. J. Chem. Soc. Perkin*, **I**, 2132–5.
DETHIER, V. G. (1941) *Amer. Naturalist*, cited by HEROUT, V., 1970.
DETROY, R. W., LILLEHOJ, E. B. & CIEGLER, A. (1971) Aflatoxin and related compounds. In KADIS, S. *et al.*, vol. VI, 4–178.
DEULOFEU, V. & RUVEDA, E. A. (1971) Venomous animals and their venom. In BÜCHERL, W. (ed.), vol. II, 475 p.
DEVYS, M. (1971) Sur la biosynthèse des stérols des végétaux. *Thèse de Doctorat. Université d'Orsay*, 90 p.
DEVYS, M., ALCAIDE, A. & BARBIER, M. (1969 a) Biosynthesis of cholesterol from pollinastanol in the tobacco *Nicotiana tabacum. Phytochem.*, **8**, 1441–4.
DEVYS, M., ALCAIDE, A., PINTE, F. & BARBIER, M. (1969 b) Pollinastanol dans la fougère *Polypodium vulgare* et la salsepareille *Smilax medica. C. R. Acad. Sc.*, **269**, 2033–5.
DEVYS, M. & BARBIER, M. (1965) Le cholestérol, principal stérol du pollen d'une composée, la porcelle *Hypochoeris radicata. C. R. Acad. Sc.*, **261**, 4901–3.
DEVYS, M. & BARBIER, M. (1967) Isolement du pollinastanol et d'une série de Δ7-stérols des pollens de deux composées. *Bull Soc. Chim. Biol.*, **49**, 865–71.
DJERASSI, C., SHIH-COLEMAN, C. & DICKMAN, J. (1974) Insect control of the future: operational and policy aspects. *Science*, **186**, 596–607.
DOSKOTCH, R. D., CHATTERJI, S. K. & PEACOCK, J. W. (1970) Elm bark derived feeding stimulants for the smaller elm bark beetle *Scolytus multistriatus. Science*, **167**, 380–2.
DOSKOTCH, R. D., MIKHAIL, A. A. & CHATTERJI, S. K. (1973) Structure of the water-soluble feeding stimulant for *Scolytus multistriatus*: a revision. *Phytochem.*, **12**, 1153–5.
DOYLE, P. J., PATTERSON, G. W., DUTKY, S. R. & THOMSON, M. J. (1972) Triparanol inhibition of sterol biosynthesis in *Chlorella emersonii. Phytochem.*, **11**, 1951–60.

DRAVNIEKS, A. (1966) Current status of odours theories. *Adv. Chem.*, **56**, 29–57.

DUBÉ, J. & LEMONDE, A. (1970) The origin of progesterone in the confused flour beetle *Tribolium confusum*. *Experientia*, **26**, 543–4.

DZUBAN, I. N. (1958) *Byull. Inst. Biol. Bodokhan Akad. Nauk URSS*, **1**, 11, cited by PIPEL, 1968.

EDGAR, J. A. & CULVENOR, C. C. J. (1975) Pyrrolizidine alkaloids in *Parsonia* species (Apocynacae) which attract danaid butterflies. *Experientia*, **31**, 393–4.

EDWARDS, P. A. & GREEN, C. (1972) Incorporation of plant sterols into membranes and its relation to sterol absorption. *FEBS letters*, **20**, 97–9.

EISNER, H. E., ALSOP, D. W. & EISNER, T. (1967) Defense mechanisms in arthropods XX: Quantitative assessment of hydrogen cyanide production in two species of millipedes. *Psyche*, **74**, 107–17.

EISNER, T. (1964) Catnip: its raison d'être. *Science*, **146**, 1318–20.

EISNER, T. (1970) *Chemical defense against predation in Arthropods*. In SONDHEIMER, E. & SIMEONE, J. B., 157–217.

EISNER, T. (1972) Chemical ecology: on arthropods and how they live as chemists. *Verhandlungsbericht der Deutschen Zool. Gesell.*, **65**, 123–37.

EISNER, T., EISNER, H. E., HURT, J. J., KAFATOS, F. C. & MEINWALD, J. (1963 a) Cyanogenic glandular apparatus of a millipede. *Science*, **139**, 1218–20.

EISNER, T., HENDRY, L. B., PEAKALL, D. B. & MEINWALD, J. (1971) 2,5-dichlorophenol (from ingested herbicide?) in defensive secretions of grasshopper. *Science*, **172**, 277–8.

EISNER, T., HURST, J. J. & MEINWALD, J. (1963 b) Defense mechanisms in arthropods XI. The structure, function and phenolic secretions of the glands of a chordeumoid millipede and a carabide beetle. *Psyche* **70**, 95–116.

EISNER, T., JOHNESSEE, J. S., CARREL, J., HENDRY, L. B. & MEINWALD, J. (1974) Defensive use by an insect of a plant resin. *Science*, **184**, 996–9.

EISNER, T., KLUGE, A. F., CARREL, J. C. & MEINWALD, J. (1972) Defense mechanisms in arthropods XXXIV. Formic acid and acyclic ketones in the spray of a caterpillar. *Ann. Entomol. Soc. Amer.*, **65**, 765–6.

EISNER, T. & MEINWALD, J. (1966) Defensive secretions of arthropods. *Science*, **153**, 1341–50.

EISNER, T., MEINWALD, J., MONRO, A. & GLENT, R. (1961) Defense mechanisms in arthropods I. The composition and function of the spray of the whip scorpion *Mastigoproctus giganteus*. *J. Insect Physiol.*, **6**, 272–8.

EISNER, T. & MEINWALD, Y. C. (1965) Defensive secretion of a caterpillar (*Papilio*). *Science*, **150**, 1733–5.

EISNER, T., PLISKE, T. E., IKEDA, M., OWEN, D. F., VASQUEZ, C., PEREZ, H., FRANCLEMONT, J. G. & MEINWALD, J. (1970) Defense mechanisms of arthropods XXVII. Osmeterial secretions of papilionid caterpillars. *Ann. Entomol. Soc. Amer.*, **63**, 914–15.

EITER, K. (1970) *Insekten-Sexuallockstoffe. Chemie der Pflanzenschutz and Schädlingsbekämpfungsmittel vol. 1.* Springer Verlag Berlin, 497–522.

ENWALL, E. L., VAN DER HELM, D., NAN HSU, I., PATTABHIRAMAN, A. J. SCHMITZ, F. J., SPRAGGINS, R. L. & WEINHEIMER, A. J. (1972) Crystal structure and absolute configuration of two cyclopropane containing marine steroids. *J. Chem. Soc. Chem. Comm.*, 215–16.

ERDMAN, T. R. & THOMSON, R. H. (1972) Sterols from the sponges *Cliona celata* and *Hymeniacidon Perleve*. *Tetrahedron*, **28**, 5163–73.

ERSPAMER, V. & ANASTASI, A. (1962) Structures and pharmacological actions of eledoisin, the active endecapeptide of the posterior salivary glands of *Eledone*. *Experientia*, **18**, 58–9.

ERSPAMER, V. & BENATI, O. (1953) Identification of murexine as beta(imidazolyl-(4)-acrylcholin. *Science*, **117**, 161–2.

ESHLEMAN, A., SIEGEL, S. M. & SIEGEL, B. Z. (1971) Is mercury from Hawaiian volcanoes a natural source of pollution? *Nature*, **233**, 471–2.

ESTABLE, C., ARDAO, M. I., BRASIL, N. P. & FIESER, L. F. (1955) Gonyleptidine. *J. Amer. Chem. Soc.*, **77**, 4942.

EUGSTER, C. H., MÜLLER, G. F. R. & GOOD, R. (1965) Wirkstoffe aus *Amanita muscaria*: Ibotensaure und Muscazone. *Tetrahedron letters,* 1813–15.

EUGSTER, C. H. & TAKEMOTO, T. (1967) Zur Nomenklatur der neuen Verbindungen aus *Amanita*-arten. *Helv. Chim. Acta,* **50,** 126–7.

EUW, J. VON, FISHELSON, L., PARSONS, J. A., REICHSTEIN, T. & ROTSCHILD, M. (1967) Cardenolides in a grasshopper feeding on milkweeds. *Nature,* **214,** 35–9.

EUW, J. VON, REICHSTEIN, T. & ROTSCHILD, M. (1969) Aristolochic acid in the swallowtail butterfly *Pachlioptera aristolochiae*. *Israel J. Chem.,* **6,** 659–70.

EVANS, D. A. & GREEN, C. L. (1973) Insect attractants of natural origin. *Chem. Soc. Rev.,* **2,** 75–97.

FERLAN, I. & LEVEZ, D. (1972) Purification and some properties of a toxin protein from *Actinia equina*. *Toxicon,* **10,** 528.

FÉVRIER, A., BARBIER, M. & SALIOT, A. (1975) Utilisation par les invertébrés marins de substances organiques en solution dans l'eau der mer: acides gras et hydrocarbures. *J. Exp. Mar. Biol. Ecol.,* (publication details n.a.).

FÉVRIER, A., BARBIER, M. & SALIOT, A. (1976) Molécules organiques dissoutes dans l'eau der mer: utilisation de l'acide palmitique, de l'alcool cétylique et du doctriacontane par les invertébrés marins. *J. Exp. Mar. Biol. Ecol.,* **25,** 123–9.

FIESER, L. F. & ARDAO, M. I. (1956) Investigation of the chemical nature of gonyleptidine. *J. Amer. Chem. Soc.,* **78,** 774–81.

FLORKIN, M. (1956) *Aspects biochimiques communs aux êtres vivants*. Masson, Paris, 425 p.

FLORKIN, M. (1966) *Aspects moléculaires de l'adaptation et de la phylogénie*. Collection GPB n° 2. Masson, Paris, 259 p.

FÖRSTER, H. & WIESE, L. (1954) Untersuchungen zur Kopulationsfäbigkeit von *Chlamydomonas eugametos*. *Z. Naturforsch.,* **9 b,** 470–1.

FUSCO, R., TRAVE, R. & VERCELLONE, A. (1955) La struttura dell'iridomyrmecina. *Chim. e Industria (Milano),* **37,** 958–60.

GALBRAITH, M. N. & HORN, D. H. S. (1966) An insect-moulting hormone from a plant. *Chem. Comm.,* 905–6.

GARY, N. E. (1974) *Behaviour and physiology of honeybees*. In BIRCH, M. C., 200–21.

GAWIENOWSKI, M. & GIBBS, C. G. (1969) Biosynthesis of pregnenolone from cholesterol in *Punicia granatum*. *Phytochem.,* **8,** 2317–19.

GEORGE, J. L. & FREAR, D. E. H. (1966) Pesticides in the Antarctic. *J. Appl. Ecol.,* **3,** Suppl. 155.

GEORGE, M. (1961) Oil pollution of marine organisms. *Nature,* **192,** 1209.

GHIRETTI, F. (1959) Cephalotoxin: the crab-paralysing agent of the posterior salivary gland of Cephalopods. *Nature,* **183,** 1192–3.

GILBERT, L. I. (1967) Lipid metabolism and function in insects. *Adv. Insect Physiol.,* **4,** 69–211.

GOAD, L. J. (1970) Natural substances formed biologically from mevalonic acid. In GOODWIN, T. W. (ed.).

GOODWIN, T. W. (ed.) (1970) *Natural substances formed biologically from MVA*. Academic Press, London, 186 p.

GOONWARDENE, H. F., WHITE, J. H., GROSVENOR, A. E. & ZEPP, D. B. (1970) Host plants and the performance of some lures for Japanese beetles. *J. Econ. Entomol.,* **63,** 1289–92.

GORDON, H. T., WATERHOUSE, D. F. & GILBY, A. R. (1963) Incorporation of ^{14}C-acetate into the scent constituents by the green vegetable bug. *Nature,* **197,** 818.

GOTO, T., KISHI, Y., TAKAHASHI, S. & HIRATA, Y. (1965) Tetrodotoxin. *Tetrahedron,* **21,** 2059–88.

GOWER, D. B. (1972) 16-unsaturated C_{19} steroids. Review of their chemistry, biochemistry and possible physiological role. *J. Steroid Biochem.,* **3,** 45–103.

GRAY, R. & BONNER, J. (1948 a) An inhibitor of plant growth from the leaves of *Encelia farinosa*. *Amer. J. Botan.,* **35,** 52–7.

GRAY, R. & BONNER, J. (1948 b) Structure determination and synthesis of a plant inhibitor, 3-acetyl 6-methoxy benzaldehyde, found in the leaves of *Encelia farinosa*. *J. Amer. Chem. Soc.,* **70,** 1249–53.

GRIMSTONE, G. (1972) Mercury in British fish. *Chem. Brit.*, **8**, 244–7.
GRÖGER, D. (1972) Ergot. In KADIS, S., *et al.*, vol. VIII 321–73.
GROSSERT, J. S. (1972) Natural products from the echinoderms. *Chem. Soc. Rev.*, **1**, 1–25.
GRÜNWALD, C. (1971) Effects of free sterols, steryl esters and steryl glycosides on membrane permeability. *Plant Physiol.*, **48**, 653–5.
GUNNISON, A. F. (1966) The collection and analysis of the volatile bee venom and solubility characteristics of bee venom solids. *PhD thesis, Cornell University, Ithaca, USA*, 55 p.
GUPTA, K. C. & SCHEUER, P. J. (1968) Echinoderm sterols. *Tetrahedron*, **24**, 5831–7.
HABERMANN, E. (1971) Chemistry, pharmacology and toxicology of bee, wasp and hornet venoms. In BÜCHERL, W. (ed.), vol. III, 61–93.
HABERMEHL, G. (1969) Chemie und biochemie von Amphibian-giften. *Naturwiss.*, **56**, 615–22.
HABERMEHL, G. (1971) *Salamander venom*. In BÜCHERL, W. (ed.), vol. II, 569–84.
HABERMEHL, G. (1975) Die biologische Bedeutung tierischer Gifte. *Naturwiss.*, **62**, 15–21.
HABERMEHL, G. & VOLKWEIN, G. (1971) Aglycones of the toxins from the cuvierian organs of *Holothuria forskali*. *Toxicon*, **9**, 319–27.
HAENNI, A. L., ROBERT, M., VETTER, W., ROUX, L., BARBIER, M. & LEDERER, E. (1965) Structure chimique des aspergillomarasmines A et B. *Helv. Chim. Acta*, **48**, 729–50.
HALSTEAD, B. W. (1965) *Poisonous and venomous marine animals of the world*. US Govern. Printing Office, Washington, vol. I, 883–965.
HALSTEAD, B. W. (1971) *Venomous coelenterates, hydroids, jellyfishes, corals and sea anemones*. In BÜCHERL, W. (ed.), vol. III, 395–416. Venomous echinoderms and annelids, starfishes, sea urchins, sea cucumbers and segmented worms, ibid., 419–41.
HAMAMURA, Y. (1965) On the feeding mechanisms and artificial food of silkworm *Bombyx mori*. *Mem. Konan Univ. Sc. Ser.*, 8 Art. **38**, 17–22.
HANS, H. & THORSTEINSON, A. J. T. (1961) The influence of physical factors and host plant odour on the induction and termination of dispersal flight in *Sitona cylindricollis*. *Ent. Exp. and Appl.*, **4**, 165–77.
HARDEGGER, E., LIECHTI, P., JACKMAN. L. M., BOLLER, A. & PLATTNER, P. A. (1963) Die Konstitution des Lycomarasmins. *Helv. Chim. Acta*, **46**, 60–74.
HARLEY, J. L. (1952) Associations between microorganisms and higher plants. *Ann. Rev. Microbiol.*, **6**, 367–86.
HASHIMOTO, Y. & YASUMOTO, T. (1960) Confirmation of saponin as a toxic principle of starfish. *Bull. Jap. Soc. Sc. Fish*, **26**, 1132–8.
HASKELL, P. T. (1961) *Insect sounds*. Quadrangle, Chicago; cited by EISNER, T., (1970).
HASTINGS, J. W., RILEY, W. H. & MASSA, J. (1965) The purification, properties and chemiluminescent quantum yield of bacterial luciferase. *J. Biol. Chem.*, **240**, 1473–81.
HEFTMANN, E. (1963) Biochemistry of plant steroids. *Ann. Rev. Plant. Physiol.*, **14**, 225–48.
HEFTMANN, E., SAUER, H. H. & BENNETT, R. D. (1968) Biosynthesis of ecdysterone from cholesterol by a plant. *Naturwiss.*, **55**, 37–8.
HENDRY, L. B., ANDERSON, M. E., JUGOVICH, J., MUMMA, R. O., ROBACKER, D. & KOSARYCH, Z. (1975) Sex pheromone of the oak leaf roller: a complex chemical messenger system identified by mass fragmentography. *Science*, **187**, 355–7.
HENZELL, R. F. & LOWE, M. D. (1970) Sex attractant of the grass grub beetle. *Science*, **168**, 1005–6.
HEROUT, V. (1970) Some relations between plants, insects and their isoprenoids. *Progress in Phytochem.*, **2**, 143–202.
HEWLINS, M. J. E., EHRHARDT, J. D., HIRTH, L. & OURISSON, G. (1969) The conversion of ^{14}C-cycloartenol and lanosterol into phytosterols by cultures of *Nicotiana tabacum*. *Europ. J. Biochem.*, **8**, 184–8.
HIKINO, H., KOHAMA, T. & TAKEMOTO, T. (1970) Biosynthesis of ponasterone A and ecdysterone from cholesterol in *Podocarpus macrophyllus*. *Phytochem.*, **9**, 367–9.

HOBSON, R. P. (1935) On a fat soluble growth factor required by blowfly larvae II. Identity with cholesterol. *Biochem. J.,* **29,** 2023–6.

HOFMANN, A., HEIM, R., BRACK, A., KOBEL, H., FREY, A., OTT, H., PETRZILKA, T. & TROXLER, F. (1959) Psilocybin und Psilocin, zwei psychotrope Wirkstoffe aus mexicanischen Rauschpilzen. *Helv. Chim. Acta,* **42,** 1557–72.

HOPKINS, C. Y., JEVANS, A. W. & BOCH, R. (1969) Occurrence of octadeca-*trans*-2, *cis*-9,*cis*-12, trienoic acid in pollen attractive to the honeybee. *Canad. J. Biochem.,* **47,** 433–6.

HOPKINS, T. A., SELIGER, H. H., WHITE, E. H. & CASS, M. W. (1967) The chemiluminescence of firefly luciferin-A model for the bioluminescent reaction and identification of the product excited state. *J. Amer. Chem. Soc.,* **89,** 7148–50.

HORN, D. H. S., MIDDLETON, E. J., WUNDERLICH, J. A. & HAMPSHIRE, F. (1966) Identity of the moulting hormones of insects and crustaceans. *Chem. Comm.,* 339–41.

HOSOZAWA, S., KATO, N. & MUNAKATA, K. (1974 a) Antifeeding substances for insects in *Caryopteris divaricata. Agr. Biol. Chem.,* **38,** 823–6.

HOSOZAWA, S., KATO, N., MUNAKATA, K. & CHEN, Y. L. (1964 b) Antifeeding active substances for insects in plants. *Agr. Biol. Chem.,* **38,** 1045–8.

HOWANITZ, W. (1959) Insects and plant galls. *Sci. Amer.,* **201,** 151–62.

HUBER, F. (1814) *Nouvelles observations sur les abeilles.* 2nd ed. translated 1926, DADANT, Hamilton, USA, 205.

HUBER, R. & HOPPE, W. (1965) Die Krystall-und Molekülstrukturanalyse des Insektenverpuppungshormons Ecdyson mit der automatisierten Faltmolekül-methode. *Chem. Ber.,* **98,** 2403–24.

HÜGEL, M.F., BARBIER, M. & LEDERER, E. (1964) Sur le pollinastanol, nouveau stérol du pollen. *Bull. Soc. Chim.,* 2012–13.

HUGHES, D. L. & COXON, D. T. (1974) Phytuberin; revised structure from X-ray crystal analysis of dihydro–phytuberin. *J.C.S. Chem. Comm.,* 822–3.

HUMM, H. J. & LANE, C. E. (1974) Bioactive compounds from the sea. *Marine Science,* vol. I, M. Decker Inc., New York, 251 p.

HUMMEL, H. & KARLSON, P. (1968) Hexansaure als Bestandteil des Spurpheromons der Termite *Zootermopsis nevadensis. Z. Physiol. Chem.,* **349,** 725–7.

HUNNEMANN, D. H. & CHRIST, V. (1974) *Phthalate esters in biological extracts.* Varian MAT, Brême, Application Note n° **17,** 1–8.

HUXLEY, J., Sir (1934) Threat and warning colouration in birds with a general discussion of the biological functions of colours. *Proceed. 8th Intern. Ornithol. Congr. Oxford,* 430–55.

IDLER, D. R., WISEMAN, P. M. & SAFE, L. M. (1970) A new marine sterol, 22-*trans*-24-norcholesta-5, 22-dien-3 β-ol. *Steroids,* **16,** 451–61.

IKAN, R., GOTTLIEB, R., BERGMANN, E. D. & ISHAY, J. (1969) The pheromone of the queen of the oriental hornet *Vespa orientalis. J. Insect Physiol.,* **15,** 1709–12.

IKEGAMI, S., KAMIYA, Y. & TAMURA, S. (1972 a) A new sterol from asterosaponins A and B. *Tetrahedron letters,* 1601–4.

Ibid. (1972 b) A novel steroid from asterosaponins. *Tetrahedron letters,* 3725–8.

IRIE, T., SUZUKI, M. & HAYAKAWA, Y. (1969) Isolation of aplysin, debromoaplysin and aplysinol from *Laurencia okamurai* Yamanda. *Bull. Chem. Soc. Japan,* **42,** 843–4.

JACOBSON, M. (1966) *Insect sex attractants.* Wiley, New York, 154 p.

JACOBSON, M. (1972) *Insect sex pheromones,* Academic Press, New York, 382 p.

JACOBSON, M., LILLY, C. E. & HARDING, C. (1968) Sex attractant of sugarbeet wire-worm: identification and biological activity. *Science,* **159,** 208–9.

JAENICKE, L. & MÜLLER, D. G. (1973) Gametenlockstoffe bei niederen Pflanzen und Tieren. *Fortsch. Chem. Org. Naturf.,* **30,** 61–100.

JAENICKE, L., MÜLLER, D. G. & MOORE, R. E. (1974) Multifidene and aucantene, C11 hydrocarbons in the male attracting scent oil from the gymnogamete of *Cutleria multifida, J. Amer. Chem. Soc.,* **96,** 3324–5.

JENNINGS, R. C., JUDY, K. J. & SCHOOLEY, D. A. (1975) Biosynthesis of the homosesquiterpenoid juvenile hormone JH II from homomevalonate in *Manduca sexta. J.C.S. Chem. Comm.,* 21–2.

JERMY, T. (1966) Feeding inhibitors and food preference in chewing phytophagous insects. *Entomol. Exptl. Appl.,* **9,** 1–12.

JOHNSON, D. F., BENNETT, R. D. & HEFTMANN, E. (1963) Cholesterol in higher plants. *Science,* **140,** 198–9.

JOHNSON, W. S., CAMPBELL, S. F., KRISHNAKUMARAN, A. & MEYER, A. S. (1969) Total synthesis of the racemic form of the second juvenile hormone from the *Cecropia* silk moth. *Proc. Nat. Acad. Sc. US.,* **62,** 1005–9.

JONES, D. A., PARSONS, J. & ROTHSCHILD, M. (1962) Release of hydrocyanic acid from crushed tissues of all stages in the life cycles of species of the *Zygenidae. Nature,* **193,** 52–3.

JUDY, K. J., SCHOOLEY, D. A., DUNHAM, L. L., HALL, M. S., BERGOT, B. J. & SIDALL, J. B. (1973) Isolation, structure and absolute configuration of a new natural insect juvenile hormone from *Manduca sexta. Proc. Nat. Acad. Sc. US.,* **70,** 1509–13.

JUSLÈN, C., WEHRLI, W. & REICHSTEIN, R. (1963) Die Glucoside des Milchsaftes von *Antiaris toxicaria* (Java). *Helv. Chim. Acta,* **46,** 117–41.

KADIS, S., CIEGLER, A. & AJL, S. J. (1971) *Microbial toxins.* Academic Press, New York, vol. VI, 563 p.

KAHANE, E. (1973) Le hasard et la vie. *Cahiers Laïques, n° 137,* 107–36.

KAHANE, E. (1974) *Lavoisier; pages choisies.* Les classiques du peuple, Paris, 271 p.

KARLSON, P. (1960) Pheromones. *Ergebniss. Biol.,* **22,** 212–25.

KARLSON, P. (1966) Ecdyson, das Hautungshormon der Insekten. *Naturwiss,* **53,** 445–53.

KARLSON, P. & Hoffmeister, H. (1963) Umwandlung von Cholesterin in Ecdyson. *Z. Physiol. Chem.,* **331,** 298–300.

KARLSON, P. & LÜSCHER, M. (1959) Pheromones: a new term for a class of biologically active substances. *Nature,* **183,** 55–6.

KARLSON, P. & SCHNEIDER, D. (1973) Sexualhormone der Schmetterlinge als Modelle chemischer Kommunikation. *Naturwiss.,* **60,** 113–21.

KARLSSON, R. & LORSMAN, D. (1972) The crystal structure of the hemihydrochloride of coccinellin, the defensive N-oxide alkaloid of the beetle *Coccinella septempunctata,* a case of symmetrical hydrogen bonding. *J. Chem. Soc. Chem. Comm.,* 627.

KASANG, G. (1973) Physicochemical processes triggered by olfaction in the silkworm moth. *Naturwiss.,* **60,** 95–101.

KATO, Y. & SCHEUER, P. J. (1974) Aplysiatoxin and debromoaplysiatoxin, constituents of the marine mollusc *Stylocheilus longicauda. J. Amer. Chem. Soc.,* **96,** 2245–6.

KATSUI, N., MATSUNAGA, A. & MASAMUNE, T. (1974) The structure of lubimin and oxylubimin, antifungal metabolites from diseased potato tubers. *Tetrahedron letters,* 4483–6.

KEM, W. R., ABBOTT, B. C. & COATES, R. M. (1971) Isolation and structure of a nemertine toxin. *Toxicon,* **9,** 15–23.

KERN, H., NAEFF-ROTH, S. & RUFFNER, F. (1972) Influence of nutritional factors on the formation of naphthazarin derivatives in *Fusarium* sp. *Phytopathol. Z.,* **74,** 272–80.

KISHI, Y., GOTO, T., INOUE, S., SUGIURA, S. & KISHIMOTO, H. (1966) *Cypridina* bioluminescence III. Total synthesis of *Cypridina* luciferin. *Tetrahedron letters* 3445–50.

KNAPP, F. F., PHILLIPS, D. O., GOAD, L. J. & GOODWIN, T. W. (1972) Isolation of 14 α-methyl-9 β 19-cyclo-5 α-ergost-24 (28)-en-3 β ol from *Musa sapientum. Phytochem.,* **11,** 3497–500.

KNOPF, J. A. E. & PITMAN, G. B. (1972) Aggregation pheromones for manipulation of the Douglas-fir beetle. *J. Econ. Entomol.,* **65,** 723–6.

KOBAYASHI, M. & MITSUHASHI, H. (1974) Structure and synthesis of amuresterol, a new marine sterol with unprecedented side chain, from *Asterias amurensis. Tetrahedron,* **30,** 2147–50. Isolation and structure of occelasterol from an annelide *Pseudoporamilla occelata. Steroids,* **24,** 399–440.

KOBAYASHI, M., TSURU, R., TODO, K. & MITSUHASHI, H. (1972) Asteroid sterols. *Tetrahedron letters,* 2935–8.

KOSUGE, T., ZENDA, H., OCHIAI, A., MASAKI, N., NOGUCHI, M., KIMURA, S. & NARITA,

H. (1972) Isolation and structure determination of a new marine toxin surrugatoxin from the Japanese ivory shell *Babylonia japonica*. *Tetrahedron letters*, 2545–8.

KULLENBERG, B. (1956) *Zool. Bidrag fr. Uppsala*, **31**, 253–4; cited by HEROUT, 1970.

LABEYRI, V. & HUIGNARD, J. (1973) Relations trophiques et comportement reproducteur des Insectes. *Ann. Soc. Roy. Zool. Belgique*, **103**, 43–51.

LANIER, G. N. & BURKHOLDER, W. E. (1974) Pheromones in speciation of Coleoptera. In BIRCH, M. C., 161–89.

LARSEN, J. B. & LANE, C. E. (1966) Some effects of *Physalia physalix* toxin on the cardiovascular system of the rat. *Toxicon*, **4**, 199–203.

LAVIE, D., JAIN, M. K. & SHPAN-GABRIELITH, S. R. (1967) A locust phagorepellent from two *Melia* species. *Chem. Comm.*, 910–11.

LAW, J. H. & REGNIER, F. E. (1971) Pheromones. *Ann. Rev. Biochem.*, **40**, 533–48.

LAW, J. H., YUAN, C. & WILLIAMS, C. M. (1966) Synthesis of a material with juvenile hormone activity. *Proc. Nat. Acad. Sc. US.*, **55**, 576–8.

LEAKE, C. D. (1968) Development of knowledge about venoms. In BÜCHERL, W. (ed.), vol, I 1–12.

LEBŒUF, M., CAVÉ, A. & GOUTAREL, R. (1964) Présence de la progestérone dans les feuilles de l'*Holarrhena floribunda. C. R. Acad. Sc.*, **259**, 3401–3.

LEDERER, E. (1950) Odeurs et parfums des animaux. *Fortshr. der Chem. Org. Naturstoffe*, **6**, 87–153.

LEDERER, E. (1969) Some problems concerning biological C-alkylation reactions and phytosterol biosynthesis. *Quarterly Rev.*, **23**, 453–81.

LEE, C. Y. (1972) Chemistry and pharmacology of polypeptide toxins in snake venoms. *Ann. Rev. Pharm.*, **12**, 265–70.

LEROY, Y. (1974) Le mimétisme animal. *La Recherche*, **5**, 417–25.

LEVITA, B. (1966) Pigmentation et photosensibilité chez *Oedipoda coerulescens. C. R. Acad. Sc.*, **262**, 2496–7.

LING, N. C., HALE, R. L. & DJERASSI, C. (1970) The structure and absolute configuration of the marine sterol gorgosterol. *J. Amer. Chem. Soc.*, **92**, 5281–2.

LLOYD, J. E. (1975) Aggressive mimicry in *Photuris* fireflies: signal repertoires by femmes fatales. *Science*, **187**, 452–3.

LOPEZ, A. & CRAWFORD, M. A. (1967) *Lancet*, **II**, 1351. Cited by DETROY, R. W., *et al.*, 1971, p. 19.

LOUSBERG, R. J. J. C. & SALEMINK, C. A. (1972) Chemistry of polysaccharide and glycopeptide phytotoxins. In WOOD, R. K. S. *et al.*, 127–37.

LOUVEAUX, J. (1965) *Plantes carnivores et végétaux hostiles*. Hachette, Paris, 108 p.

LUCOTTE, G. (1974) Polychromatisme de la coquille de l'œuf de la caille domestique *Coturnix coturnix japonica* I. Classification des phénotypes à l'intérieur de la forme dominante. *Bull. Soc. Biol.*, **168**, 422–4.

LUCOTTE, G., CHOUSSY, M. & BARBIER, M. (1974) Polychromatisme de la coquille de l'œuf de la caille domestique *Coturnix coturnix japonica*. II. Isolement et dosage des pigments des phénotypes à l'intérieur de la forme dominante. *C. R. Soc. Biol.*, **168**, 425–8.

MCCAPRA, F. (1973) Chimie de la bioluminescence, *Endeavour (Fr. edn)*, **32**, 139–45.

MCCAPRA, F., CHANG, Y. C. & FRANCIS, V. P. (1968) The chemiluminescence of a firefly luciferin analogue. *Chem. Comm.*, 22–3.

MCCLINTOCK, M. K. (1971) Menstrual synchrony and suppression. *Nature*, **229**, 244–5.

MCELROY, W. D. & STREHLER, B. L. (1954) Bioluminescence. *Bacteriol. Rev.*, **18**, 177–94.

MCMICHAEL, D. F. (1971) Molluscs, classification, distribution, venom apparatus, and venoms, symptomatology of stings. In BÜCHERL, W. (ed.), vol. III, Academic Press, London, 373–93.

MACALPINE, G. A., RAPHAEL, R. A., SHAW, A., TAYLOR, A. W. & WILD, H. J. (1974) Synthesis of the germination stimulant ± strigol. *J.C.S. Chem. Comm.*, 834–5.

MACHLIS, L., NUTTING, W. H. & RAPOPORT, H. (1968) The structure of sirenin. *J. Amer. Chem. Soc.*, **90**, 1674–6.

MACHLIS, L., NUTTING, W. H., WILLIAMS, W. H. & RAPOPORT, H. (1966) Production, isolation and characteristization of sirenin. *Biochemistry*, **5**, 2147–52.

112 References

MacMorris, T. C. & Barksdale, A. W. (1967) Isolation of sex hormone from the water mold *Achlya bisexualis*. *Nature*, **215**, 320–1.

Mann, T. (1931) *La montagne magique*. Arthème Fayard, Paris, vol. II, p. 323.

Manser, P., Slàma, K. & Landa, V. (1968) Sexually spread insect sterility induced by the analogues of juvenile hormone. *Nature*, **219**, 395–6.

Maretic, Z. (1971) Latrodectism in Mediterranean countries, including south Russia, Israel and North Africa. In Bücherl, W. (ed.), vol. III, 299–310.

Martin, P. & Rademacher, B. (1960) Studies on the mutual influences of weeds and crops. *Brit. Ecol. Soc. Symp.*, **1**, 143–52.

Maschwitz, U. W. (1964) Alarm pheromones and alarm behaviour in social Hymenoptera. *Nature*, **204**, 324–7.

Mathias, A. P., Ross, D. M. & Schachter, M. (1960) The distribution of 5-hydroxy-tryptamine and other substances in sea anemones. *J. Physiol. London*, **151**, 296–311.

Matsuda, H., Tomiie, Y., Yamamura, & Hirata, Y. (1967) The structure of Aplysin-20. *Chem. Comm.*, 898–9.

Matsumoto, Y. (1970) In *Control of insect behaviour by natural products*. Wood, D. L., Silverstein, R. M. & Nakajima, M. (eds), Academic Press, New York, p. 189.

Matsumoto, T., Yanaguja, M., Maeno, S. & Yasuda, S. (1968) A revised structure of pederine. *Tetrahedron letters*, 6297–300.

Matsumura, F., Coppel, H. C. & Tai, A. (1968) Isolation and identification of termite trail-following pheromone. *Nature*, **219**, 963–4.

Mayer, F. L., Stalling, D. L. & Johnson, J. L. (1972) Phthalate esters as environmental contaminants. *Nature*, **238**, 411–13.

Mayo, P. de, Spencer, E. Y. & White, R. W. (1963) Terpenoids IV. The structure and stereochemistry of helminthosporal. *Canad. J. Chem.*, **41**, 2996–3004.

Mebs, D. (1973) Chemistry of animal venoms, poisons and toxins, *Experientia*, **29**, 1328–34.

Meinwald, J., Chadha, M. S., Hurst, J. J. & Eisner, T. (1962) Defense mechanisms of arthropods IX. Anisomorphal, the secretion of a phasmid insect. *Tetrahedron letters*, 29–33.

Meinwald, J., Erickson, K., Hartshorn, M., Meinwald, Y. C. & Eisner, T. (1968 a) Defensive mechanisms of arthropods XXIII. An allenic sesquiterpenoid from the grasshopper *Romalea microptera*. *Tetrahedron letters*, 2959–62.

Meinwald, J., Meinwald, Y. C., Chalmers, A. M. & Eisner, T. (1968 b) Dihydromatricaria acid: acetylenic acid secreted by soldier beetles. *Science*, **160**, 890–2.

Meinwald, J., Happ, G. M., Labows, J. & Eisner, T. (1966) Cyclopentanoid terpene biosynthesis in a phasmid insect and in catmint. *Science*, **151**, 79–80.

Meinwald, J., Meinwald, Y. C. & Mazzocchi, P. H. (1969) Sex pheromone of the queen butterfly: chemistry. *Science*, **164**, 1174–5.

Meinwald, J., Opheim, K. & Eisner, T. (1972) Gyridinal: a sesquiterpenoid aldehyde from the defensive glands of gyrinid beetles. *Proc. Nat. Acad. Sc., US*, **69**, 1208–10.

Meinwald, J., Thomson, W. R., Eisner, T. & Owen, D. F. (1971) Pheromones VII. African monarch: major components of the hair pencil secretion. *Tetrahedron letters*, 3485–8.

Meinwald, Y. C., Meinwald, J. & Eisner, T. (1966) 1, 2-dialkyl-4 (3 H)-quinazolinones in the defensive secretions of a millipede (*Glomeris marginata*). *Science*, **154**, 390–1.

Melrose, D. R., Reed, H. C. B. & Patterson, R. L. S. (1971) *Br. Vet. J.*, **127**, 197; cited by Brooksbank *et al.*, 1974.

Menn, J. J. & Beroza, M. (1972) *Insect juvenile hormones; chemistry and action*. Academic Press, New York, 341 p.

Meyer, A. S., Schneiderman, A. A., Hanzmann, E. & Ko, J. H. (1968) The two juvenile hormones from the *Cecropia* silk moth. *Proc. Nat. Acad. Sc. US*, **60**, 853–60.

Meyer, K. & Linde, H. (1971) Collection of toad venoms and chemistry of the toad venom steroids. In Bücherl, W. (ed.), vol. II, 521–56.

MICHAEL, R. P., BONSALL, R. W. & WARNER, P. (1974) Human vaginal secretions: volatile fatty acid content. *Science*, **186**, 1217–19.

MICHAEL, R. P. & KEVERNE, E. B. (1970) Primate sex pheromones of vaginal origin. *Nature*, **225**, 84–5.

MICHAEL, R. P. & SAAYMAN, G. (1967) Sexual performance index of male rhesus monkeys. *Nature*, **214**, 425.

MILLER, L. P. & FLEMION, F. (1973) The role of minerals in phytochemistry. *Phytochemistry*, vol. III, Van Nostrand-Reinhold, New York, 1–40.

MILLS, A. L. (1971) Lead in the environment. *Chem. Brit.*, **7**, 160–2.

MIRANDA, F., KUPEYAN, C., ROCHAT, H., ROCHAT, C. & LISSITZKY, S. (1970) Isolation and characterization of eleven neurotoxins from the venoms of the scorpions *Androctonus australis*, *Buthus occitanus tunetanus* and *Leiurus quinquestriatus quinquestriatus*. *Eur. J. Biochem.*, **16**, 514–23.

MIRANDA, F. & LISSITZKY, S. (1958) Purification de la toxine du venin de scorpion *Androctonus australis*. *Biochem. Biophys. Acta*, **30**, 217–18.

MIRANDA, F. & LISSITZKY, S. (1961) Scorpamins: the toxic proteins of scorpion venom. *Nature*, **190**, 443–4.

MIROCHA, C. J., CHRISTENSEN, C. M. & NELSON, G. H. (1971) F-2 (zearalenone) estrogenic mycotoxin from *Fusarium*. In KADIS, S. *et al.*, vol. VII, 107–38.

MOORE, B. P. (1968) Studies on the chemical composition and function of the cephalic gland secretion in Australian termites. *J. Insect Physiol.*, **14**, 33–9.

MOORE, B. P. (1974) Pheromones in the termite societies. In BIRCH, M. C., 250–66.

MOORE, R. E. & SCHEUER, P. J. (1971) Palytoxin: a new marine toxin from a coelenterate. *Science*, **172**, 495–8.

MOREAU, C. (1974) *Moisissures toxiques dans l'alimentation*. Masson, Paris, 471 p.

MORISAKI, M., OHATAKA, H., AWATA, N., IKEKAWA, N., HORIE, Y. & NAKASONE, S. (1974) Nutritional effects of possible intermediates of phytosterol dealkylation in the silk worm *Bombyx mori*. *Steroids*, **24**, 165–76.

MORSE, R. A. & BOCH, R. (1971) Pheromone concert in swarming honeybees. *Ann. Entomol. Soc. Amer.*, **64**, 1414–17.

MORSE, R. A., SHEARER, D. A., BOCH, R. & BENTON, A. W. (1967) Observations on alarm substances in the genus *Apis*. *J. Apicult. Res.*, **6**, 113–18.

MOSHER, H. S., FUHRMAN, F. A., BUCHWALD, H. D. & FISCHER, H. G. (1964) Tarichatoxin: tetrodotoxin, a potent neurotoxin. *Science*, **144**, 1100–10.

MOSS, M. O. (1971) Rubratoxins, toxic metabolites of *Penicillium rubrum*. In KADIS, S. *et al.*, vol. VI, 381–407.

MÜLLER, C. H. & DEL MORAL, R., (1966) Soil toxicity induced by terpenes from *Salvia leucophylla*. *Bull. Torrey Botan. Club*, **93**, 130–7.

MÜLLER, C. H., JAENICKE, L., DANIKE, M. & AKINTOBI, R. (1971) Sex attractant in a brown alga: chemical structure. *Science*, **171**, 815–16.

MÜLLER, D. G. & JAENICKE, L. (1973) Fucoserraten, the female sex attractant of *Fucus serratus*. *FEBS letters*, **30**, 137–9.

MÜLLER-SCHWARTZE, D. (1969) Complexity and relative specificity in a mammalian pheromone. *Nature*, **223**, 525–6.

MÜLLER-SCHWARTZE, D. (1974) Mammalian pheromones: active components in the subauricular scent of the male pronghorn. *Science*, **183**, 860–2.

MUTO, T. & SUGAWARA, R. (1965) The housefly attractants in mushrooms. I. Extraction and activities of the attractive components in *Amanita muscaria*. *Agric. Biol. Chem.*, **29**, 949–55.

MUTO, T. & SUGAWARA, R. (1970) In *Control of insect behaviour by natural products*. WOOD, D. L., SILVERSTEIN, R. M. & NAKIJAMA, M. (eds), Academic Press, New York, 133.

NAKANISHI, K. (1971) The ecdysones. *Pure Appl. Chem.*, **25**, 167–95.

NAKANISHI, K., KOREEDA, M., SASAKI, S., CHANG, M. L. & HSU, H. Y. (1966) Insect hormones. The structure of ponasterone A, an insect moulting hormone from the leaves of *Podocarpus nakaii*. *Chem. Comm.*, 915–17.

NASON, A. & MACELROY, W. D. (1963) Modes of action of the essential mineral elements. In STEWART, F. C. (ed.), *Plant Physiol.*, vol. III, Academic Press, New York, 451–536.

NAYAR, J. K. & FRAENKEL, G. (1963) The chemical basis of host selection in the catalpa sphinx *Ceratomia catalpae. Ann. Entomol. Soc. Amer.*, **56**, 119–22.

NISHIDA, R., FUKUMI, H. & ISHII, S. (1974) Sex pheromone of the German cockroach *Blatella germanica*, responsible for male wing rising. *Experientia*, **30**, 978–9.

NIXON, A. C. (1972) *Autoxidation et antioxydants*. Interscience, New York, vol. II, 695 p.

OKAISHI, T. & HASHIMOTO, Y. (1962 a) Physiological activities of nereistoxine. *Bull. Jap. Soc. Sc. Fish.*, **28**, 930–5.

OKAISHI, T. & HASHIMOTO, Y. (1962 b) The structure of nereistoxine. *Agr. Biol. Chem.*, **26**, 224–7.

OPARIN, A. (1962) *L'origine et l'évolution de la vie*. Éditions de la paix, Moscou, 224 p.

OPPELZER, E., PRELOG, V. & SENSI, P. (1964) Konstitution des Rifamycins B und Verwandter Rifamycine. *Experientia*, **20**, 336–9.

OWENS, L. D., THOMSON, J. F., PITCHER, R. G. & WILLIAMS, T. (1972) Structure of rhizobitoxine, an antimetabolic enol-ether amino-acid. *Chem. Comm.*, 714.

PAIN, J. & BARBIER, M. (1963) Structures chimiques et propriétés biologiques de quelques substances identifiées chez l'abeille. *Insectes Sociaux*, **10**, 129–42.

PAIN, J., BARBIER, M. & BOGDANOVSKY, D. (1962) Chemistry and biological activity of the secretions of queen and worker honeybees. *Comp. Bioch. Physiol.*, **6**, 233–41.

PALLARES, E. S. (1946) Note on poisonous principle of *Polydesmus vicinus. Arch. Biochem.*, **9**, 105–8.

PATTERSON, R. L. S. (1968) Identification of 3 α-hydroxy-5 α androst-16 ene as the musk odor component of boar submaxillary salivary gland and its relation to the sex odor taint in pork meat. *J. Sc. Food Agric.*, **19**, 434–8.

PAVAN, M. (1959) Biochemical aspects of insect poisons. *Proc. 4th Cong. Biochim.*, **12**, 15–36.

PAVAN, M. (1975) Gli iridoidi negli insetti. *Publicazioni dell Instituto di Entomologia Agraria dell Universita di Pavia*, **2**, 1–49.

PAVAN, M. & TRAVE, R. (1958) Étude sur les *Formicidae* IV. Sur le venin du *Dolichoderidae Tapinoma nigerrimum. Insectes Sociaux*, **5**, 299–308.

PAYNE, T. L., SHOREY, H. H. & GASTON, L. K. (1973) Sex pheromones of Lepidoptera XXXVIII. Electroantennogram response in *Autographa california* to *cis*-7 dodecenylacetate and related compounds. *Ann. Entomol. Soc. Amer.*, **66**, 703–4.

PERCY, J. E. & WEATHERSTON, J. (1974) Gland structure and pheromone production in insects. In BIRCH, M. C., 11–61.

PERISSE, A. C. M. & SALES, C. A. (1970) Estudo quimico de diplopoda brasileiros IV. *Orthoporus fuscipes. An. Acad. Sc. Brasil. Cienc.*, **42**, Suppl. 199–204.

PERUN, T. J. & EGAN, R. S. (1969) The conformation of erythromycin aglycones, *Tetrahedron letters*, 387–90.

PESCE, H. & DELGADO, A. (1971) Poisoning from adult moths and caterpillars. In BÜCHERL, W. (ed.), vol. III, 119–56.

PFEIFFER, W. (1974) Pheromones in fish and amphibians. In BIRCH, M. C., *Pheromones*, 269–96.

PHISALIX, M. (1922) *Animaux venimeux et venins*. Vol. I, Masson, Paris, 656 p.

PICARELLI, Z. P. & VALLE, J. R. (1971) Pharmacological studies on caterpillars venoms. In BÜCHERL, W. (ed.), vol. III, 103–8.

PILPEL, N. (1968) Le sort du pétrole à la surface de la mer. *Endeavour (Fr. edn)*, **27**, 11–12.

PONSINET, G. & OURISSON, G. (1967) Biosynthèse *in vitro* des triterpènes dans le latex d'*Euphorbia. Phytochem.*, **6**, 1235–43.

PONSINET, G. & OURISSON, G. (1968) Études chimiotaxonomiques dans la famille des Euphorbiacées III. Répartition des triterpènes dans le latex d'*Euphorbia. Phytochem.*, **7**, 89–98.

POPPLESTONE, C. R. & UNRAU, A. M. (1974) Studies on the biosynthesis of antheridiol. *Canad. J. Chem.*, **52**, 462–8.

POULTON, E. B. (1888) The secretion of pure formic acid by lepidopterous larvae for the purpose of defence. *Brit. Assoc. Adv. Sc. Rep.,* **5,** 765–6.
POURNELLE, G. H. (1968) The *Platypus* and its venomous characteristics. In BÜCHERL, W. (ed.), vol. I, 31–41.
PREMUZIC, E. (1971) Chemistry of natural products derived from marine sources. *Fortsch. Chem. Org. Naturstoffe,* **29,** 417–88.
PUCEK, M., (1968) Chemistry and pharmacology of insectivores' venoms. In BÜCHERL, W. (ed.), vol. I, 43–50.
QUILICO, A., PIOZZI, F. & PAVAN, M. (1957) The structure of dendrolasine. *Tetrahedron,* **1,** 177–85.
RADEMACHER, B. (1941) Uber den antagonistischer Einfluss von Roggen und Weizen auf Keimung und Entwicklung mancher Unkraüter. *Pflanzenbau,* **17,** 131–43.
RAHN, R. (1968) Rôle de la plante-hôte sur l'attractivité sexuelle chez *Acrolepia assectella. C. R. Acad, Sc., Paris,* **266,** 2004–6.
RALLS, K. (1971) Mammalian scent marking. *Science,* **171,** 443–9.
RAMADE, F. (1974) *Éléments d'écologie appliquée.* Édiscience, Paris, 522 p.
RAPER, J. R. (1952) Chemical regulation of sexual processes in the Thallophytes. *Botan. Rev.,* **18,** 447–546.
REES, C. J. C. (1969) Chemoreceptor specificity associated with choice of feeding site by the beetle *Chrysolina brunsvicensis* on its food plant *Hypericus hirsutum. Entomol. Exp. Appl.,* **12,** 565–83.
REGNIER, F. E. & WILSON, E. O. (1968) The alarm defense system of the ant *Acanthomyops claviger. J. Insect Physiol.,* **14,** 955–70.
REICHSTEIN, T. (1967) Cardenolide als Abwehrstoffe bei Insekten. *Naturwiss. Rundchau,* **20,** 499–511.
REICHSTEIN, T., VON EUW, J., PARSONS, J. A. & ROTHSCHILD, M. (1968) Heart poisons in the monarch butterfly. *Science,* **161,** 861–6.
RENWICK, J. A. A. & VITE, J. P. (1970) Symposium on population attractants. Contribution from the Boyce-Thomson Institute, **24,** n° 13, 283.
RESCHKE, T. (1969) Die Gamone aus *Blakeslea trispora. Tetrahedron letters,* 3435–9.
RICE, R. D. & HALSTEAD, B. W. (1968) Report of fatal cone shell sting by *Conus geographus. Toxicon,* **5,** 223–4.
RICHE, C., PASCARD-BILLY, C., DEVYS, M., GAUDEMER, A., BARBIER, M. & BOUSQUET, J. F. (1974) Structure cristalline et moléculaire de la phoménone, phytotoxine produite par le champignon *Phoma exigua var. non oxydabilis. Tetrahedron letters,* 2765–6.
RIDDIFORD, L. M. & WILLIAMS, L. M. (1967) Volatile principle from oak leaves: role in sex life of *Polyphemus* moth. *Science,* **158,** 139–41.
RITTER, F. J. & WIENTJENS, W. H. J. M. (1967) Sterol metabolism of insects. *TNO Nieuws,* **22,** 381–92.
ROBERTS, J. C. (1974) Aflatoxins and sterigmatocystins. *Fortschr. Chem. Org. Naturstoffe,* **31,** 119–51.
ROBBINS, W. E., DUTKY, R. C., MONROE, R. E. & KAPLANIS, J. N. (1962) The metabolism of H^3-β-sitosterol by the German cockroach. *J. Entomol. Soc. Amer.,* **55,** 102–4.
ROCHAT, H., ROCHAT, C., MIRANDA, F. & LISSITZKY, S. (1967) Presence of an identical amino-acid sequence in the neurotoxins of *Androctonus australis. B. B. Res. Comm.,* **29,** 107–12.
ROCHAT, C., SAMPIERI, F., ROCHAT, H. & MIRANDA, F. (1972) Iodination of neurotoxins I and II of the scorpion *Androctonus australis. Biochimie,* **54,** 445–9.
RODIN, J. O., SILVERSTEIN, R. M., BURKHOLDER, W. E. & GORMAN, J. E. (1969) Sex attractant of female dermestid beetle *Trogoderma inclusum. Science,* **165,** 904–5.
ROELOFS, W. L. & CARDE, R. T. (1974) Sex pheromones in lepidopterous species. In BIRCH, M. C., 96–114.
ROELOFS, W. L. & COMEAU, A. (1971) Sex pheromone perception synergist and inhibitors for the red-banded leaf roller attractant. *J. Insect Physiol.,* **17,** 435–48.
ROHMER, M. & BRANDT, R. D. (1973) Les stérols et leurs précurseurs chez *Astasia longa. Europ. J. Biochem.,* **36,** 446–54.

ROLLER, P., DJERASSI, C., CLOETENS, R. & TURSCH, B. (1969) The isolation of three new holothurinogenins and their chemical correlation with lanosterol. *J. Amer. Chem. Soc.,* **91,** 4918–20.

ROMANUK, M., SLÀMA, K. & SORM, F. (1967) Constitution of a compound with a pronounced juvenile hormone activity. *Proc. Nat. Acad. Sc. US.,* **57,** 349–58.

ROPARTZ, P. (1966) Contribution à l'étude du déterminisme d'un effet de groupe chez les souris. *C. R. Acad. Sc.,* **262,** 2070–2.

ROPARTZ, P. (1968) Rôle des communications olfactives dans le comportement social des souris mâles. *Colloque Intern. CNRS, Paris,* **173,** 323–39.

ROSENFELD, I. & BEATH, O. A. (1964) *Selenium geobotany, biochemistry, toxicity and nutrition,* Academic Press, New York, p. 61.

ROTBERG, A. (1971) Lepidopterism Brazil. In BÜCHERL, W. (ed.), vol. III, 157–68.

ROTHWEILER, W. & TAM, C. (1970) Isolierung und Struktur der Antibiotica Phomin und 5-dehydrophomin. *Helvita Ch. Acata,* **53,** 696–724.

RÜDIGER, W., KLOSE, W., VUILLAUME, M. & BARBIER, M. (1968) On the structure of pterobilin, the blue pigment of *Pieris brassicae. Experientia,* **24,** 1000.

RÜDIGER, W., KLOSE, W., VUILLAUME, M. & BARBIER, M. (1969) Sur la biliverdine IX α, pigment vert des insectes Orthoptères *Clitumnus extradentatus* et *Mantis religiosa. Bull. Soc. Chim. Biol.,* **51,** 559–62.

RUSSELL, BERTRAND (1962) *Ma conception du Monde;* Gallimard, Paris, 183 p.

RUSSELL, D. W. (1966) Cyclodepsipeptides. *Quarterly Rev.,* **20,** 559–76.

SAITO, M., ENOMOTO, M. & TATSUNO, T. (1971) Yellowed rice toxins, luteoskyrin and related compounds, chlorine-containing compounds and citrinin. In KADIS, S. *et al.,* vol. VI, 299–380.

SAKAI, T., MURAI, F., BUTZUGAN, Y. & SUZUI, S. (1959) On the structure of actinidine and matatabilactone, the effective component of *Actinidia polygama. Bull. Chem. Soc. Jap.,* **32,** 315–16.

SAKAI, T., FUJINO, A., MURAI, F., SUZUI, A. & BUTZUGAN, Y. (1959) The structure of matatabilactone. *Bull. Chem. Soc. Jap.,* **32,** 1154–55.

SAKAI, T., NISHIMURA, K. & HIROSE, Y. (1965) The structure and stereochemistry of four new sesquiterpenes isolated from the wood oil of kaya *Torreya nucifera. Bull. Chem. Soc. Jap.,* **38,** 381–7.

SALIOT, A. & BARBIER, M. (1971) Sur l'isolement de la progestérone et de quelques céto-stéroïdes de la partie femelle des gonades de la coquille St Jacques *Pecten maximus. Biochimie,* **53,** 265–6.

SALIOT, A. & BARBIER, M. (1974) Acides gras, hydrocarbures et stérols dissous dans l'eau de mer. *Congr. Intern. Geochim., Actes,* 607–17.

SANCRIN, E. & BOISSONNAS, R. A. (1962) Synthesis of eledoisin. *Experientia,* **18,** 59–61.

SAUER, H. H., BENNETT, R. D. & HEFTMANN, E. (1968) Ecdysterone biosynthesis in *Podocarpus elata. Phytochem.,* **7,** 2027–30.

SCHANTZ, E. J., LYNCH, J. M., VAYVADA, G., MATSUMOTO, K. & RAPOPORT, H. (1966) The purification and characterization of the poison produced by *Gonyaulax catenella* in axenic culture. *Biochem.,* **5,** 1191–5.

SCHANTZ, E. J., MOLD, J. D., STRANGER, D. W., SHAVEL, J., RIEL, F. J., BOWDEN, J. P., LYNCH, J. M., WYLER, R. S., RIEGEL, B. & SOMMER, E. (1957) Paralytic shellfish poison VI. A procedure for the isolation and purification of the poison from the toxic clam and mussel tissues. *J. Amer. Chem. Soc.,* **79,** 5230–5.

SCHANTZ, E. J., GHAZAROSSIAN, V. E., SCHNOES, H. K., STRONG, F. M., SPRINGER, J. P., PEZZANITE, J. O. & CLARDY, J. (1975) Structure of saxitoxin. *J. Amer. Chem. Soc.,* **97,** 1238–9.

SCHENBERG, S. & PEREIRA-LIMA, F. A. (1971) *Phoneutria nigriventer* venom. Pharmacology and biochemistry of its components. In BÜCHERL, W. (ed.), vol. III. 279–97.

SCHEUER, P. J. (1971) Toxins from marine invertebrates. *Naturwiss.,* **58,** 549–54.

SCHEUER, P. J. (1973) *Chemistry of marine natural products.* Academic Press, London, 201 p.

SCHILDKNECHT, H. (1970) The defensive chemistry of land and water beetles. *Angew. Chem. Intern. ed.,* **9,** 1–9.

SCHILDKNECKT, H. (1971) Sommets d'évolution dans la chimie défensive des insectes. *Endeavour (Fr. edn)*, **30**, 136–41.

SCHILDKNECHT, H. & HOLOUBEK, K. (1961) Die Bombardierkäfer und ihre Explosionschemie. *Angew. Chem.*, **73**, 1–7.

SCHILDKNECHT, H. & RAUCH, G. (1961) Die chemische Natur der Luftphytoncide von Blattpflanzen insbezondere von *Robinia pseudacacia. Z. Naturforsch.*, **16 b**, 422–9.

SCHILDKNECHT, H. & SCHMIDT, H. (1963) Die chemische zusammensetzung des Wehrsekretes von *Dicranura vinula. Z. Naturforsch.*, **186**, 585–7.

SCHILDKNECHT, H., TAUSCHER, B. & KRAUSS, D. (1972) Der Duftstoffe des Traumelkäfer *Gyrinus natator. Chem. Zeit., chem. App.* **96**, 33–5.

SCHILDKNECHT, H. & WEISS, K. H. (1961) Chinone als aktives Prinzip der Abwehrstoffe von Diplopoden. *Z. Naturforsch.*, **16 b**, 810–16.

SCHILDKNECHT, H., WENNEIS, W. F., WEIS, K. H. & MASCHWITZ, U. M. (1966) Glomerin, ein neues Arthropoden-Alkaloid. *Z. Naturforsch.*, **21**, 121–7.

SCHMIALEK, P. (1961) Die Identifisierung zweier im Tenebriokot und in Hefe Vorkommender Substanzen mit juvenilhomone Wirkung. *Z. Naturforsch.*, **16 b**, 461–4.

SCHNEIDER, D. (1957) Electrophysiological investigations on the antennal receptor of the silk moth during chemical and mechanical stimulation. *Experientia*, **31**, 89–91.

SCHNEIDER, D. (1962) Electrophysiological investigations on the olfactory specificity of sexual attracting substances in different species of moths. *J. Insect Physiol.*, **8**, 15–30.

SCHNEIDER, D., KASANG, G. & KAISSLING, G. (1968) Bestimmung der Riechschwelle von *Bombyx mori* mit Tritium-markierten Bombykol. *Naturwiss.*, **55**, 395.

SCHOOLEY, D. A., JUDY, K. J., BERGOT, B. J., HALL, M. S. & SIDALL, J. B. (1973) Biosynthesis of the juvenile hormones of *Manduca sexta*: labelling pattern from mevalonate, propionate and acetate. *Proc. Nat. Acad. Sc., US.*, **70**, 2921–5.

SEIGLER, D. S. (1975) Naturally occuring cyanogenic compounds. *Phytochem.*, **14**, 9–29.

SHEIKH, Y. M., TURSCH, B. M. & DJERASSI, C. (1972) 5 α-pregn-9 (11) ene-3 β, 6α-diol-20-one and 5 α-cholesta-9 (11), 20 (22) diene-3 β, 6 α-diol-23-23-one. *J. Amer. Chem. Soc.*, **94**, 3278–80.

SHIMOMURA, O. & JOHNSON, F. H. (1968) The structure of *Latia* luciferin. *Biochem.*, **7**, 1734–8. Purification and properties of the luciferase and of a protein cofactor in the bioluminescence system of *Latia neritoides*. Ibid., 2574–80.

SHIMOMURA, O. & JOHNSON, F. H. (1971) Mechanism of luminescent oxidation of *Cypridina* luciferin. *Biochem. Biophys. Res. Comm.*, **44**, 340–6.

SHIMOMURA, O. & JOHNSON, F. H. (1972) The structure of the light emitting moiety on Aequorin. *Biochem.*, **11**, 1602–8.

SHIMOMURA, O., JOHNSON, F. H. & KOHAMA, Y. (1972) Reactions involved in bioluminescence systems of limpet *Latia neritoides* and luminous bacteria. *Proc. Nat. Acad. Sc., US.*, **69**, 2086–9.

SHOREY, H. H., GASTON, L. K. & JEFFERSON, R. N. (1968) Insect sex pheromones. In *Advances in Pest Control Research.* MEDCALF, R. (ed.), Interscience, New York, 57–126.

SHRIFT, A. (1969) *Ann. Rev. Plant Physiol.*, **20**, 475–94; cited by MILLER & FLEMION, 1973.

SILVERSTEIN, R. M., RODIN, J. O., BUKHOLDER, W. E. & GORMAN, J. E. (1967) Sex attractant of the black carpet beetle. *Science*, **157**, 85–6.

SILVERSTEIN, R. M. (1970) Attractant pheromones in Coleoptera in chemicals controlling insect behaviour. In BEROZA, M. (ed.), Academic Press, New York, 21–40.

SINGH, H., MOORE, R. E. & SCHEUER, R. J. (1967) The distribution of quinone pigments in echinoderms. *Experientia*, **23**, 624–6.

SODERQUIST, C. J. (1973) Juglone and allelopathy. *J. Chem. Ed.*, **50**, 782–3.

SOLEIL, J. & LALLOZ, L. (1971) Les psychodysleptiques. *Prod. Probl. Pharm.*, **26**, 682–90.

SONDHEIMER, E. & SIMEONE, J. B. (1970) *Chemical Ecology.* Academic Press, New York, 336 p.

Souza, J. J. de, Ghisalberti, E. L., Rees, J. H. & Goodwin, T. W. (1970) Studies on insect moulting hormones: biosynthesis of ecdysone, ecdysterone and 5 β-hydroxy ecdysterone in *Polypodium vulgare. Phytochem.*, **9**, 1247–52.

Standifer, L. N., Devys, M. & Barbier, M. (1968) Pollen sterols: a mass spectrographic survey. *Phytochem.*, **7**, 1361–5.

Stein, G. (1963) Untersuchungen uber den sexuallockstofe der Hummelmänchen. *Biol. Ztbl.*, **82**, 343–9.

Stewart, W. W. (1971) Isolation and proof of structure of wildfire toxin. *Nature*, **229**, 174–8.

Steyn, P. S. (1971) Ochratoxin and other dihydroisocoumarins. In Kadis, S. *et al.*, vol. VI, 179–205.

Suzuki, A., Taguchi, H. & Tamura, S. (1970) Isolation and structure elucidation of three new insecticidal cyclodepsipeptides, destruxins C and D and desmethyldestruxin B produced by *Metarrhizium anisophae. Agric. Biol. Chem. Jap.*, **34**, 813–16.

Suzuki, N., Sato, M., Nishikawa, N. & Goto, T. (1969) Synthesis and spectral properties of 2-(6'-hydroxybenzothiazol-2'-yl)-4-hydroxythiazole, a possible emitting species in the firefly bioluminescence. *Tetrahedron letters*, 4683–4.

Svoboda, J. A. & Robbins, W. E. (1967) Conversion of β-sitosterol to cholesterol blocked in an insect by hypocholesterolemic agents. *Science*, **154**, 1131–2.

Tai, A., Matsumura, F. & Coppel, H. C. (1969) Chemical identification of the trail following pheromone for a southern subterranean termite. *J. Org. Chem.*, **34**, 2180–2.

Tam, N. D. (1970) La guerre chimique. *La Recherche* n° **5**, 442–54.

Taylor, A. (1971) Toxicology of sporidesmins and of the epipolythiadioxopiperazines. In Kadis, S. *et al.*, vol. VII, 337–76.

Templeton, G. E., Meyer, W. L., Grable, C. I. & Sigel, C. W. (1967) The chlorosis toxin from *Alternaria tenuis* is a cyclic tetrapeptide. *Phytopathol.*, **57**, 833.

Tette, J. P. (1974) Insect population management. In Birch, M. C., 399–410.

Thiessen, D. D., Regnier, F. E., Rice, M., Goodwin, M., Isaacks, N. & Lauson, N. (1974) Identification of a ventral scent marking pheromone in the male Mongolian gerbil *Meriones unguiculatus. Science*, **184**, 83–5.

Thorsteinston, A. J. (1953) The chemotactic responses that determine host specificity in an oligophagous insect *Plutella maculipennis. Canad. J. Zool.*, **31**, 52–72.

Tokuyama, T., Daly, J. & Witkop, B. (1972) The structure of batrachotoxin, a steroidal alkaloid from the Colombian arrow poison frog *Phyllobates aurotaenia* and partial synthesis of batrachotoxin and its analogs and homologs. *J. Amer. Chem. Soc.*, **91**, 3931–8.

Tokuyama, T., Uenoyama, K., Brown, G., Daly, J. W. & Witkop, B. (1974) Allenic and acetylenic spiropiperidine alkaloids from the neotropical frog. *Dendrobates histrionicus. Helvetica Chim. Acta*, **57**, 2597–604.

Tschesche, R. (1965) Les stéroïdes en C21 d'origine végétale. *Bull. Soc. Chim. Fr.*, 1219–28.

Tschesche, R. & Lillienweiss, G. (1964) Cardenolid-Biosynthese aus Pregnenolon Glucosid. *Z. Naturforsch.*, **196**, 265–6.

Tsuda, K. (1966) Uber tetrodotoxin, Giftstoff der Bowlfische. *Naturwiss.*, **53**, 171–6.

Tumlinson, J. H., Hardee, D. D., Gueldner, R. C., Thompson, A. C., Hedin, P. A. & Minyard, J. P. (1969) Sex pheromones produced by male boll weevil: isolation, identification and synthesis. *Science*, **166**, 1010–12.

Tumlinson, J. H., Moser, J. C., Silverstein, R. M., Brownlee, R. G. & Ruth, J. M. (1972) A volatile trail pheromone of the leaf-cutting ant *Atta texana. J. Insect Physiol.*, **18**, 809–15.

Turner, A. B., Smith, D. S. H. & Mackie, A. M. (1971) Characterization of the principal steroid saponins of the starfish *Marthasterias glacialis*: structure of the aglycones. *Nature*, **233**, 209–10.

Tursch, B., Cloetens, R. & Djerassi, C. (1970) Chemical studies of marine invertebrates VI. Terpenoids LXV. Praslinogenin, a new holothurinogenin from the Indian Ocean sea cucumber *Bohadschia koellikeri. Tetrahedron letters*, 467–70.

Tursch, B., Daloze, D., Dupont, M., Pasteels, J. M. & Tricot, M. C. (1971) A defense alkaloid in a carnivorous beetle. *Experientia*, **27**, 1380–1.

US Bureau of Commercial Fisheries. Surface slicks have 10,000 times more pesticides than encircling water. *Commerc. Fish Rev.*, **32**, n° 7.

VAN HOOGT, E. G. (1936) Aseptic cultures of insects in vitamin research. *Z. Vitaminforsch.*, **5**, 118–26.

VEITH, J. J., BARBIER, M., PAIN, J. & ROGER, B. (1974) Transformation de steroïdes par l'abeille *Apis mellifica*. *Comp. Biochem. Physiol.*, **47**, 459–72.

VELGOVA, H., CERNY, V., SORM, F. & SLÀMA, K. (1969) Further compounds with antisclerorization effect on *Pyrrhocoris apterus* larvae. Structure and activity correlations. *Coll. Cz. Chem. Comm.*, **34**, 3354–76.

VERRON, H. & BARBIER, M. (1962) L'héxène-3 ol-1 substance attractive des termites *Calotermes flavicollis* et *Microcerotermes endentatus*. *C. R. Acad. Sc.*, **254**, 4089–91.

VIALA, J., DEVYS, M. & BARBIER, M. (1972) Sur la structure des stérols à 26 atomes de carbone du Tunicier *Halocynthia roretzi*. *Bull. Soc. Chim. Fr.*, 3626–7.

VIGNY, A. & MICHELSON, A. M. (1974) Studies in bioluminescence XIII. Bioluminescence bactérienne: mise en evidence et identification du produit de transformation de l'aldéhyde. *Biochimie*, **56**, 171–6.

VOGT, W. (1970) What is a toxin? *Toxicon*, **8**, 251.

VOOGT, P. A. (1972) Investigation of the capacity of synthesizing 3 β sterols in mollusca VI. The biosynthesis and composition of 3 β sterols in the neogasteropods *Purpura lappillus* and *Murex brandaris*. *Comp. Biochem. Physiol.*, **47**, 459–72.

VUILLAUME, M. (1968) Pigmentation et variation pigmentaire de trois Insectes: *Mantis religiosa*, *Sphodromantis viridis* et *Locusta migratoria*. *Bull. Biol.*, **102**, 148–232.

VUILLAUME, M. (1969) *Les pigments des Invertébrés*. Masson, Paris, 184 p.

VUILLAUME, M., CHOUSSY, M. & BARBIER, M. (1970) Pigments tétrapyrroliques verts et bleus des ailes de Lépidoptères. Ptérobilines et néoptérobilines. *Bull. Soc. Zool. Fr.*, **95**, 19–28.

VRKOC, J. & UBIK, K. (1974) 1-nitro *tràns*-1 pentadecene as the defensive compound of termites. *Tetrahedron letters*, 1463–4.

WADA, K., ENOMOTO, Y., MATSUI, K. & MUNAKATA, K. (1968 a) Insect antifeedants from *Parabenzoin trilobum*: two new sesquiterpenes, shiromodiol diacetate and monoacetate. *Tetrahedron letters*, 4673–6.

WADA, K. & MUNAKATA, K. (1968 b) Insect antifeedants from *Parabenzoin trilobum*, absolute configuration of shiromodiol diacetate. *Tetrahedron letters*, 4677–80.

WALDNER, E. E., SCHATTER, C. & SCHMIDT, H. (1969) Zur Biosynthese des Dendrolasins, eines Inhaltstoffes der Ameise *Lasius fuliginosus*. *Helv. Chim. Acta*, **52**, 15–24.

WASSERMAN, R. H., CORRADINO, R. A. & KROOK, L. P. (1975) *Cestrum diurnum*: a domestic plant with 1, 25-dihydroxycholecalciferol-like activity. *Biochem. Biophys. Res. Comm.*, **62**, 85–91.

WATKINS, J. F., COLE, T. W. & BALDRIDGE, R. S. (1967) Laboratory studies on interspecies trail following and trail preference of army ants. *J. Kansas Entomol. Soc.*, **40**, 146–51.

WEATHERSTON, J. (1967) The chemistry of arthropods' defensive substances. *Quarterly Rev.*, **21**, 287–313.

WEATHERSTON, J. & MACLEAN, W. (1974) The occurence of (E)-11-tetradecen-1-ol, a known sex attractant inhibitor in the abdominal tips of virgin female eastern spruce budworm *Choristoneura fumiferana*. *Canad. Entomol.*, **106**, 281–4.

WEAVER, N., WEAVER, C. C. & LAW, J. H. (1964) The attractiveness of citral to foraging honeybees. *Progr. Rep. Tex. Agric. Exptl. Stn.*, n° **2324**, 1–7.

WEBB, L. J., TRACEY, J. G. & HAYDOCK, K. P. (1967) A factor toxic to seedlings of the same species associated with living roots of the non gregarious subtropical rain forest tree *Grevillea robusta*. *J. Appl. Eccl.*, **4**, 13–25.

WECKERING, R. (1960) Stéréoélectronie de poisons de fourmis et de coléoptères. *Proc. 11th Cong. Entomol.*, **3**, 102–9.

WENSLER, R. J. D. (1962) Mode of host selection by an aphid. *Nature*, **195**, 830–1.

WENT, F. W. (1964) The nature of aitken condensation nuclei in the atmosphere. *Proc. Nat. Acad. Sc. US*, **51**, 1259–67.

120   *References*

WHEELER, J. W., EVANS, S. L., BLUM, M. S. & TORGERSON, R. L. (1975) Cyclopentylketones: identification and function in *Azteca* ants. *Acience*, **187**, 254–5.
WHEELER, J. W., HURST, J. J., MEINWALD, J. & EISNER, T. (1964) *Trans*-2 dodecanal and 2-methyl-1, 4-quinone produced by a millipede. *Science*, **144**, 540–1.
WHEELER, J. W., VON ENDT, D. W. & WEMMER, C. (1975) 5-thiomethyl pentane-2, 3 dione. A unique natural product from the striped hyena. *J. Amer. Chem. Soc.*, **97**, 441–2.
WHITE, E. H., MACCAPRA, F. & FIELD, G. F. (1963) The structure and synthesis of firefly luciferin. *J. Amer. Chem. Soc.*, **85**, 337–43.
WHITTAKER, V. P. (1959) The identity of natural and synthetic ββ-dimethyl acrylylcholine. *Biochem. J.*, **71**, 32–8.
WHITTAKER, V. P. (1960) Pharmacologically active choline esters in marine gastropods. *Ann. NY. Acad. Sc.*, **90**, 695–705.
WHITTAKER, R. H. (1970 a) The biochemical ecology of higher plants. In SONDHEIMER, E. & SIMEONE, J. B., 1970, 43–70.
WHITTAKER, R. H. (1970 b) *Communities and ecosystems*. Macmillan, New York, 210 p.
WHITTAKER, R. H. & FEENY, P. P. (1971) Allelochemics: chemical interactions between species. *Science*, **171**, 757–70.
WHITTEN, W. K. (1966) Pheromones and mammalian reproduction. *Advances Reprod. Physiol.*, **1**, 155–7.
WHITTEN, W. K. & CHAMPLIN, A. K. (1973) The role of olfaction in mammalian reproduction. *Endocrinology* II, Part *1, Handbook of Physiology*; 103–23.
WICKLER, W. (1968) *Le mimétisme animal et végétal*. Coll. l'Univers des Connaissances. Hachette éd., Paris, 254 p.
WIELAND, T. & WIELAND, O. (1971) The toxic peptides of *Amanita* species. In KADIS, S. *et al.*, vol. VIII, 249–80.
WIESE, L. & JONES, R. F. (1963) Studies on gamete copulation in heterothallic chlamydomonads. *J. Cellular Comp. Physiol.*, **61**, 265–74.
WILKINSON, S. (1961) The history and chemistry of muscarine. *Quarterly Rev.*, **15**, 153–71.
WILLIAMS, C. M. & LAW, J. H. (1965) The juvenile hormone IV. Its extraction, assay and purification. *J. Insect Physiol.*, **11**, 569–80.
WILLIAMS, C. M. & SLÀMA, K. (1966) The juvenile hormone VI. Effects of the 'paper factor' on the growth and metamorphosis of the bug *Pyrrhocoris apterus*. *Biol. Bull.*, **130**, 247–53.
WILSON, E. O. (1962) Pheromones. *Sci. Amer.*, **208**, 100–14.
WILSON, E. O. (1965) Chemical communication in the social insects. *Science*, **149**, 1064–71.
WILSON, R. E. & RICE, E. L. (1968) Allelopathy as expressed by *Helianthus annuus* and its role in old field succession. *Bull. Torry Botan. Club*, **95**, 432–8.
WONG, J. L., OESTERLIN, R. & RAPOPORT, H. (1971) The structure of saxitoxin. *J. Amer. Chem. Soc.*, **93**, 7344–5.
WOOD, R. K. S., BALLIO, A. & GRANITI, A. (1972) *Phytotoxins in plant diseases*. Academic Press, London, 530 p.
WOODRUFF, H. B. & MACDANIEL, L. E. (1958) In *The strategy of chemotherapy*. COWAN, S. T. & ROWATT, E. (eds), Cambridge Univ. Press, p. 29.
WOODWARD, R. B. (1964) The structure of tetrodotoxin. *Pure Appl. Chem.*, **9**, 49–74.
WURSTER, S. F. (1968) DDT reduces photosynthesis by marine phytoplankton. *Science*, **159**, 1474–5.
YANG, C. C., YANG, H. J. & HUANG, J. S. (1969) The amino-acid sequence of cobrotoxin. *Biophys. Biochem. Acta*, **188**, 65–77.
YINON, V., SHULOV, A. & IKAN, R. (1971) The olfactory responses of granary beetles towards natural and synthetic fatty acid esters. *J. Insect Physiol.*, **17**, 1037–49.
ZANDEE, D. I. (1966) Metabolism in the crayfish *Astacus astacus* III. Absence of cholesterol synthesis. *Arch. Intern. Physiol. Biochem.*, **74**, 435–41.
ZO BELL, C. E. (1964) *Advances in water pollution research*. Pergamon Press, Oxford, p. 85.
ZLOTKIN, E. (1973) The chemistry of animal venoms. *Experientia*, **29**, 1453–66.

Index

acetic acid: in ant venom, 32; in arachnid defensive fluid, 29
acetylcholine, in hornet venom, 28
acetylenic acid, defensive substance of soldier beetle, 36–7
Achlya (slime mould), steroid sex pheromone of, 61, 77–8
acrylylcholine, in toxin of whelk, 25
Actinomycetes, antibiotics from, 21
adrenaline, noradrenaline: in toad venoms, 38
aerosols, plant products in form of, 10
aflatoxins, of fungi, 12, 13–14
aggregation pheromones, 73–5
alarm pheromones: of fish, 81; of insects, 70–2
aldehydes, aliphatic: in bacterial luminescence, 45–6
algae, sex pheromones of, 78
alkaloids: of ladybirds, 36, 37, 43; of plants, accumulated by insects, 31; of plants, precursors of insect pheromones, 67; in skin of frogs, 37–8
allelochemical (interspecific) interactions, 4, 21
allelopathy, 7
allomones, 3, 4, 25, 28
allyl isothiocyanate of Cruciferae: repellant for many insect species, attractant for *Pieris,* 9
alternaric acid, mycotoxin, 20
Amanita muscaria, fly attractant of, 83
Amanita spp., toxins of, 15–17
amatoxins, amanitins, of *Amanita* (cyclic octapeptides), 15, 16
amines, in invertebrate toxins, 25
amino acids: in luciferins, 45; in marasmins, linked by bonds other than peptide, 18
D-amino acids: in firefly luciferin, 44; in peptide antibiotics, 24
amino-glycoside antibiotics, 21–2
amphibians, venoms of, 37–8
amygdalin, cyanogenetic glucoside, 7

anabasein, annelid toxin, 25
anisomycin, antibiotic active against protozoa, 24
annelids, marine: toxins of, 25
antagonistic action, of substances co-existing with pheromones, 67
antamide, substance in *Amanita* acting against *Amanita* toxins, 15, 16
antennae of insects, measurement of electric currents in (electroantennograms), 64
antheridiol, steroid sex pheromone of *Achlya,* 61, 77–8
antibiotics, 13, 21–5; and over-population, 93
ants: alarm pheromones of, 70–2; trail pheromones of, 72–3; venoms of, 31–3
apamin, peptide in bee venom, 28
aphids, alarm pheromones of, 72
aphrodisiacs, pheromones acting as, 64, 67, 68
aplysin, aplysinol, aplysiatoxin: toxins of *Aplysia* spp., also present in algae, 25–6
aposematism (warning colours): in insects, 41–3; in mammals, 80
arachnids, chemical defence in, 28–9
arbutin (hydroquinone glycoside), growth inhibitor, 7
aristolochic acid; in plants, and in caterpillars feeding on them, 31
Armilla mellea, luminescent fungus, 45
arrow poisons, 30, 37
aspergillomarasmins, fungal phytotoxins, 18
Aspergillus spp.: aflatoxins of, 13, 14; production of penicillins by, 24
association of plants, 9
asterosaponins, of echinoderms, 26, 52
Astragalus spp., accumulation of selenium in, 11
ATP, in luminescence of fireflies, 44
attractants for insects, in plants, 83–4
aureomycin A, polycyclic antibiotic, 24